STATISTICAL
METHODS
for Librarians

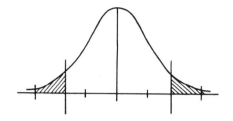

STATISTICAL
METHODS
for Librarians

RAY L. CARPENTER
*University of North Carolina
at Chapel Hill*

with

ELLEN STOREY VASU
*University of North Carolina
at Chapel Hill*

AMERICAN LIBRARY ASSOCIATION

Chicago 1978

Acknowledgment is made to the literary executor of the late Sir Ronald A. Fisher, F.R.S., to Dr. Frank Yates, F.R.S., and to Longman Group Ltd., London, for permission to reprint tables 3, 4, and 7 from their book *Statistical Tables for Biological, Agricultural, and Medical Research* (6th edition, 1974).

Library of Congress Cataloging in Publication Data

Carpenter, Ray L   , 1926–
  Statistical methods for librarians.

  Bibliography: p.
  Includes index.
  1. Statistics. 2. Library statistics. I. Vasu,
Ellen Storey, 1948–   II. Title.
HA29.C297   519.5'02'4092   78–3476
ISBN 0–8389–0256–1

                                         Printed in the United States of America

Contents

Introduction, ix

Chapter 1. **DESCRIPTIVE STATISTICS**   1
Levels of Measurement, 2; Nominal, 2; Ordinal, 3;
Interval, 4; Ratio, 5; Summarizing Statistics, 6;
Proportion, 6; Percentage, 7; Ratios and Rates, 8;
Summarizing Measures, 9; Central Tendency, 13;
Measures of Dispersion, 16; Coefficient of Variation,
19; The Normal Distribution, 21; Z Scores, 24;
Z Score Sample Problem, 25; Chebyshev's Theorem,
25; Exercises, 26; References and Suggested
Readings, 27

Chapter 2. **SAMPLING**   28
Probability Sampling, 30; Simple Random Sampling,
30; Systematic Sampling, 32; Stratified Sampling, 34;
Correcting Disproportionately Stratified Samples, 36;
Nonprobability Sampling, 37; Purposive Sampling,
37; Quota Sampling, 38; Accidental Sampling, 38;
Sample Size, 39; References and Suggested Readings,
40

Chapter 3. **INDUCTIVE STATISTICS**   41
The Statistical Hypothesis, 42; Hypothesis Testing
about the Population Mean, 44; The Central-Limit

Theorem, 48; Testing a Hypothesis about μ When
n Is Large, 53; Testing a Hypothesis about μ When
n Is Small, 56; Confidence Limits, 59; Difference of
Means Test, 60; Exercise, 64; References and
Suggested Readings, 64

Chapter 4. **CORRELATION AND REGRESSION**   65
Correlation, 65; Scattergrams, 66; Calculating r, 70;
Hypothesis Tests about the Population Correlation,
71; Regression Analysis, 72; Exercises, 76; References and Suggested Readings, 76

Chapter 5. **NONPARAMETRIC TESTS AND MEASURES**   77
Nominal Scale Tests and Measures, 78; The Chi-
Square Test of Independence, 78; Strength of
Association, 85; The Contingency Coefficient, 86;
Ordinal Scale Tests and Measures, 87; The Wald-
Wolfowitz Runs Test, 87; Spearman's Rank-Order
Correlation, r_s, 89; Use of r_s with Interval Scale
Data, 90; Kendall's *Tau*, 91; Exercise, 94; Refer-
ences and Suggested Readings, 94

Appendix Tables
1. Table of Random Digits, 96–97
2. Areas under the Normal Curve, 98–99
3. Distribution of t, 100–1
4. Distribution of χ^2, 102–3
5. Values of the Correlation Coefficient, 104–5
6. Critical Values of R, 106

Glossary, 107

Answers to Exercises, 115

Index, 117

FIGURES
1. Frequency Polygon Displaying Stops for 17 Counties   12
2. Histogram   12
3. Frequency Polygon Displaying Staff Salaries for 9
   Librarians   15
4. Frequency Polygon Displaying Staff Salaries   16
5. A Normal Distribution with Population Mean Equal to 25 22

6. A Normal Distribution, Indicating 1 Standard Deviation
   Unit above and below the Mean   23
7. A Normal Distribution, Indicating Placement of 1, 2, and
   3 Standard Deviation Units on Either Side of the Mean   23
8. A Multimodal Distribution   26
9. Illustration of Proportional Stratified Random Sampling   34
10. Theoretical Frequency Distribution for Total Population
    of Library Users on Variable Number of Books Borrowed   47
11. Theoretical Sampling Distribution of the Mean Number of
    Books Borrowed by Library Users   47
12. Theoretical Sampling Distribution of $\overline{X}$ 51
13. Z Score Transformation from a Sampling Distribution   52
16. Sampling Distribution   54
15. Student's t Distribution   58
14. Unit Normal Curve   61
17. Scattergram Containing Data Points for 10 Libraries
    on 2 Variables   67
18. Scattergram Showing Relationship between 2 Variables   68
19. Scattergram Illustrating No Association between Variables
    X and Y 69
20. Scattergram Illustrating Nonrepresentative Sample   69
21. Chi-Square Distribution   82

TABLES

1. Statistical Measures and Tests for Data Analysis   3
2. Data on Library Collections   7
3. Scheduled Stops of Bookmobiles   10
4. Frequency Distribution of Scheduled Stops   10
5. Scheduled Stops for 17 Counties   11
6. Frequency Distribution of Stops for 17 Counties   11
7. Apex Branch Circulation Figures   17
8. Number of Professional Personnel   20
9. Types of Sampling Techniques under Consideration   29
10. Data for 500 Adults on Library Usage and Graduate Status   36
11. Distribution of Number of Books Borrowed   46
12. Data for 10 Branch Libraries Measured on 2 Variables   67
13. Data Gathered from 10 Branch Libraries on 2 Variables   71
14. Data for 11 Cases on Variables X and Y 75
15. Cross Tabulation of Sex by Library Usage   80

Introduction

The research process is a special and powerful mode of thinking. It yields an understanding of how related events behave, and will often provide ways of coping with the environment. It comprises all the methods of gathering numerical data and drawing conclusions about their characteristics and their behavior. The kind of data selected for study is determined by past developments in the field in which these data belong. The actual methods used in arranging them and drawing the conclusions belong to statistics.

In many respects the value of knowing how research is done is greater for librarians, archivists, and information scientists than for most other professionals. To understand why this is so, it is useful to explore several propositions about the nature of research and the role of the librarian. (For the sake of simplicity, the term librarian will include information scientists, archivists, and documentalists.)

As the world grows ever more complex and its parts more interdependent, certain kinds of organizations have emerged that are of particular interest to us. These are the organizations that act as developers, processors, and disseminators of information. Within this context, the librarian's uses for research are threefold. As mediators between recorded information and users, librarians must be able not only to locate information but also to interpret or evaluate this information for patrons. Much information is in the form of or is based on research monographs, articles, or reports which the librarian must first identify and select and then be able to disseminate. By understanding both the language and the general principles, as well as the methods that make up this literature, the librarian can fill his or her role intelligently. Thus knowledge of the research process helps the librarian provide more useful, more critical, and more balanced information, and one of the profession's primary functions is greatly enriched.

Two other dimensions of the librarian's role are rewarded similarly. First, the librarian is a consumer of various data, and studies them in order to better his or her professional performance or the services of the organization, a point we will turn to in a moment. Second, although the number may now be modest, librarians will be increasingly expected to be participants in research projects. Such projects may be mainly concerned with library or information-centered activities; however, librarians can also be involved in other kinds of studies. Public and school librarians may participate in community educational studies that embrace a wide range of social, economic, and political and technological factors. Academic and special librarians may be involved in similar studies in their own organizational settings, and they may participate in the research activities of particular disciplines.

As a consumer of research, the librarian finds the literature of his or her field increasingly statistical. The journals, books, and other resources from relevant fields, such as business administration, sociology, or political science, are likely to be even more quantitative. As active researchers, either as resource persons or as principal investigators, librarians are virtually bound to handle numerical data. Consequently, this book is devoted to selected procedures of data analysis commonly known as social science statistics.

As mentioned above, librarians spend considerable time studying research in the literature of librarianship and other fields and are increasingly expected to devote energy to research on their organizations. This has come about because of the growing importance of research in management and the growing importance of management in the role of the librarian.

Research yields new findings. These are critical to operating large-scale analyses of library or information networks or of individual libraries or information centers. That is, the very basis of rational management depends on information. The goal of research is to provide information that is reliable and valid, so that decision making by librarians and by those whom librarians serve can be optimally beneficial. The quality of the data (information) derived from research determines the importance of its contribution to the effectiveness of library service and to the effectiveness of those whom librarians serve. It is hoped the knowledge of the limits, as well as the benefits of statistical analysis, will improve the quality of management information.

In recent years, political, social, and economic conditions have generated a related factor known as public accountability. In the past two decades, support and need for libraries has increased to the point that funding agencies are compelled to require precise rationalization of library operations. Laissez-faire attitudes toward library management have rapidly declined as the visibility of the library, largely due to its greater economic resources, has increased. Public accountability is not peculiar to libraries, of course; it is a growing expectation of nearly all organizations. At the same time, the economic problems of libraries have become greater in kind and in scope. Costs of materials and personnel have risen and new and costly methods of bibliographic and circulation control have been introduced, to cite but two examples. Resolving these problems is greatly enhanced by the use of research, not only for accountability but also to help meet a fundamental professional goal: the constant improvement of service in the face of ever-changing demands and resources to implement the goal. In any event, the ritualistic prescription or ascription of the virtues of libraries, without substantial rationalization based on careful qualitative and quantitative analysis, is likely to fall on deaf ears among those who provide support for libraries.

Two major factors, then, can be seen as compelling in improving the ability of librarians to evaluate and to conduct research: (1) complications that are brought about by changes in communication and technology and (2) public and professional expectation of accountability in quantitative terms. Pressures from general social and technological conditions and from governments and parent institutions apart, there is increasing evidence that larger numbers of librarians are determined better to understand their professional world and its context. Such librarians find many statistical procedures helpful and apply them accordingly.

Research, if it is reasonably valid, gives us a basis for critical appraisal or evaluation. This point merits closer scrutiny. First, it is assumed in many quarters that research is not much good without a formally stated conceptual (theoretical) orientation or origin. According to this notion, a researcher confronts a body of theory, or at least some kind of abstract concept, and turns to empirical study for illumination, "proof," or description of this concept. This assumption might better read: Abstract concepts should be tested empirically. To do this requires a certain mode of reasoning, as well as the selection of appropriate procedures.

But another test of the reality of this world shows that research is not necessarily or simply grounded on formal concepts. As previously stated, the research process is a special way of *coping*, of adapting and adjusting to our environment. The process may at times begin with a neat abstract formulation, or it *may* begin with a naive, exploratory hunch, or it may begin with both. Indeed, the value of the techniques and results of research is not simply that it permits us more careful examination of reality than does intuition or abstraction alone. There is great value in the research process itself. The research process can sharpen our conceptualizations and improve our rationalizations (explanations) *if* we are obedient to its results and the general credo of the relentlessly curious and intellectually vigorous scientist or scholar.

The very technology of research and statistical analysis provides us with a framework for conceptualizing ourselves and our world, including the work and service in libraries, that abstraction alone could not.

The search for fact and for truth is essential, and there are many starting points in the search. It is one thing to acknowledge the need to test our concepts with study and the need to make the design of that study efficient by clearly conceptualizing it at the onset. It is another to recognize that research can be legitimately based on loosely formulated hunches, although the design of the research must be conceptually clear. The distinction is important; to confuse the issue leads to excessive formalism in research ("It's no good if there isn't a theory behind it") and to sloppy and invalid research ("Investigating a great hunch shouldn't be inhibited by a lot of methodological strictures").

Thus research can and should be guided by clear and concise theories or concepts, insofar as possible. However, research that is based on minimal conceptualization can produce useful information that may, in turn, lead to redefining and clarifying conceptual understanding. Librarianship, at this point in time, lacks a highly developed systematic conceptual framework for explaining its various purposes and functions. Many librarians and other scholars and scientists are working to develop theories and laws of library behavior. In addition, within the numerous concepts and theories that embrace the studies of human behavior, management science, and information science is a reservoir of largely untapped ideas that, with little reformulation, could be used fruitfully for research, to the end of developing intellectual foundations for the field. Further, as the study of similar institutions, particularly other public service organizations, progresses, librarians will be able to borrow ever more heavily for hypotheses and assumptions to test, reframed to fit the particular characteristics of libraries and information centers.

Already, reading and applying findings in relevant disciplines, as well as in library and information science, can yield a store of researchable ideas that is large and complex enough to fill (if not exceed) any present capacity in the field.

In the following chapters of this book we hope to give a basic understanding of statistics, statistical analysis, and its usefulness in library science. In attempting this, we have kept the mathematical prerequisites at a minimal level. Our goal is not to turn our readers into accomplished statisticians, which we could not do in a few chapters. Our goal is to acquaint readers with the most common statistical techniques and terminology encountered in any basic research area. After reading this text and attempting the problems at the end of each chapter, one should have a fundamental grasp of some of the basic elements of statistical analysis. For those who would like to explore statistical and methodological topics beyond the scope of this text, we have included references at the end of each chapter.

Our discussion begins in chapter 1 with an introduction to descriptive statistics that are geared toward quantifying characteristics of a group of individuals or cases on a number of variables of interest to us. In order to quantify these characteristics, we must learn something about the measurement process which, essentially, assigns numbers to individuals according to the qualities or amount of the variable they possess. This presentation is also included in the first chapter.

Chapter 1 also contains the methods for calculating certain summary statistics. Through them, we are able to summarize great amounts of information by means of a few statistics or numbers. Sometimes, however, we are not interested in only the small number of individuals or cases to which we have immediate access. Rather, we would like to make generalizations about a larger number of individuals (i.e., all library users), based on our small subgroup of library users on whom we have data. To make this inference valid, we must become acquainted with some of the ways to acquire such a subgroup or sample in order to make sounder generalizations.

Chapter 2 deals with samples and sampling procedures. Based on our newly acquired knowledge, we turn to inferential statistics in chapter 3 and cover some of the fundamental concepts on which inductive statistics are based. It is at this point that we will begin using statistical methods for making generalizations and testing our statistical hypothesis in a logical inductive manner.

In chapter 4 we expand on inductive techniques and begin to examine relationships between variables that are of interest to us. Measures of association are presented that help us summarize information about the form and strength of the relationships we find among our variables. All of the techniques which are discussed up to this point are of a powerful nature, which requires rather strict assumptions about our data and data-gathering methods.

The final topic (chapter 5), which is extremely useful, is elementary nonparametric methods. Through knowledge of them, we should be able to add a certain amount of flexibility to our research capabilities.

Throughout the whole presentation, examples are given, based on questions of interest to librarians. Interpretations of the findings are included to unify the research question process. At the end of each chapter we have added simple exercises and answers, so that one can immediately begin applying what one learns. In addition, in a glossary at the end of the text we redefine and explain some of the essential concepts in this text.

We hope that our ideas will stimulate your interest in research and statistics, and provide you with new ways of conceptualizing, formulating, and answering questions that are of interest to you as a professional in the field.

1
Descriptive
Statistics

There are two broad classes of statistics, descriptive and inferential. *Descriptive* statistics enable one to summarize a large body of data, consisting of one or more variables, to make them intelligible. A *variable* is any item, quality, operation, or other phenomenon that one wants to analyze. It is understood to be subject to change or variation, hence a "variable." For any statistical analysis, variables must of course in some way be operational and measurable.

For example, if we are interested in the use of a library, one operational and measurable variable that can be selected to analyze library use is the circulation of materials—assuming that records of circulation are available or can be developed. If you have a list of the daily circulation figures for a year, you will need to perform some kind of operation to make the numbers manageable and meaningful. The process of doing so is a kind of summarizing, such as finding an average daily circulation for the year. At this descriptive level we restrict our interest and generalizations to the case or the body of data at hand.

When we have only a part of the entire body of data about one or more variables, we may employ *inferential* or *inductive* statistical procedures. In short, we may have only a sample of all daily circulation figures and want to make some general statements about the circulation. To obtain valid results, it is necessary to meet specific requirements that go beyond those of descriptive measurements. These requirements are quite logical and straightforward. They will be discussed in the section devoted to inductive or inferential statistics.

Levels of Measurement

We have been using the term measure without being very specific, as if it were unambiguous. Often there is no problem with the kind of measurement we take, especially if we are working with standardized scales or units of measurement, such as feet and inches or the metric scale. However, many variables that are of great importance to librarianship are not susceptible to measurement on so refined a level. Statistical procedures, nevertheless, are based on mathematical reasoning that includes assumptions about the level of measurement.

We will not present the mathematical foundations of the statistical procedures that are used in this book but we must, nonetheless, understand that the requirements of the specific tests and other procedures have built-in assumptions which must be met. Table 1 shows which of the statistical tests and measures covered in this text are appropriate for use with each of the four *levels of measurement*: nominal, ordinal, interval, and ratio.

Nominal

To any librarian or information scientist the *nominal* level of measurement will hardly be new. Indeed, the basic analytical scheme of any science is nominal, which is to say that it is classificatory. Separating things by type would seem to be rudimentary and obvious, but it is nonetheless extremely important—and, like many products of man's knowledge, seems simple only after the system or typology is produced.

The Dewey Decimal system is an example of a nominal level of measurement: it distinguishes among books or other materials only as they fit into one class or another. The library public also may be measured nominally, that is, categorized or classified in many ways; for instance, as users or nonusers of libraries. Among users we may have further subcategories, such as male and female or blue collar and white collar. Within a library staff, we often classify members as professional, paraprofessional, and clerical.

Such classifications may, of course, be an underlying part or first step of any level of measurement, but they can be extremely useful in themselves. Searching for and finding common factors within and among the various categories can tell us a great deal. Are there more males or females among library users than among nonusers? More college graduates than high school graduates?

In each category there are real numbers, of course, and these numbers can be manipulated in many ways *within* each category. The problem

Table 1. **Statistical Measures and Tests for Data Analysis**

Nominal	
One Variable	*Two Variables*
Calculation of:	Chi-square hypothesis test of
Proportions	independence (χ^2)
Percentages	Calculation of:
Ratios	Contingency coefficient (C)—
Rate(s) of change	measure of association
Mode	Tschuprow's T—measure of
	association
	Cramer's V—measure of
	association

Ordinal	
One Variable	*Two Variables*
Mode	Wald-Wolfowitz runs test
	Calculation of:
	Spearman rank-order correlation
	coefficient (r_s)
	Kendall's *tau* measure of
	association (τ)

Ratio or Interval	
One Variable	*Two Variables*
Hypothesis test on	Hypothesis test on:
population mean (t test)	Population correlation
Calculation of:	coefficient (ρ)
Mean	Difference between 2 population
Median	means (t test)
Mode	Calculation of:
Mean deviation	Pearson product–moment correlation
Standard deviation	coefficient (r)
Coefficient of	Simple regression equation
variability	

of level of measurement occurs when one mistakenly assumes that he or she may treat the categories as ordered or continuous when they are, in fact, discrete and qualitative. Discussion of the other levels of measurement—and of the special techniques for analyzing nominal data—will make the issue clearer.

Ordinal

An *ordinal* scale enables us to distinguish relationships in terms of rank-order. A given library may be "greater than" or "less than" any other library or libraries in terms of certain variables. The collection of library A may be larger than that of library B, which is larger than that of library C. Ordinal measurement permits us to make more powerful statements than does the nominal scale. We may not only determine that a given collection is typologically distinctive but also that it may be compared with other libraries' collections, at least in relative terms of rank.

Feelings of preference are frequently measured in ordinal terms. A library user may be asked to rank-order the importance of six major services offered by a library. The user, we understand, cannot say with any greater precision how he or she values these services.

For instance, in rating reference services, the book collection, and the record-tape-film collection the user may rank the book collection first (most important), record-tape-film second, and reference services third. We know only the relative position of these services; there is no known "distance" between the first and second and the second and third choices.

Consider another example. What are your favorite ten novels? You may rank *War and Peace* overwhelmingly as number one—so far ahead of the other nine that the distance is greater from first to second than from second to tenth. Very often people have a distinctive first choice, then favor two or three titles more or less equally, then rank others rather clearly below and in sequence, with little feeling about the magnitude of distinctions among them.

Interval

An *interval* scale not only allows us to categorize and rank observations for a variable, it also gives us a fixed unit of measurement on the scale and an arbitrary zero point. The fixed unit means that the distance between any two adjacent points on the scale is the same.

For example, the variable IQ is usually considered to be measured at the interval level. This means that the distance between 90 and 100 measures the same amount of IQ as the distance between 110 and 120. Also, an IQ score of 0 does not mean that the individual has no intelligence, since the scale for IQ was arbitrarily selected. A person of average intelligence could just as easily have been assigned a score of 500, rather than 100. The numbers are interpretable only in terms of the IQ instrument which was used to measure the concept in the first place.

Few variables of interest to librarians are measured at the interval level.

Ratio

Most of us take for granted, as "the" type of measurement, the *ratio* scale, where the distances between points on the scale are precise and known and there is an absolute zero. Units on such a scale are usually fixed and standardized, allowing us to perform much more rigorous and precise analysis.

If branch library A has 5,000 square feet of floor space, branch B has 4,000, and branch C has 2,000, we know not only that branch B is smaller than A and larger than C (i.e., we can rank-order the branches by size of floor space), but we can also state exactly how many feet, or percentages, larger and smaller it is. Branch A is 1,000 square feet, or 25 percent, larger than branch B. A is 3,000 square feet, or 150 percent, larger than branch C. Branch B is 2,000 square feet, or 50 percent, larger than branch C. A value of 0 means that there is absolutely no floor space; it indicates total absence of the property being measured, hence the term absolute zero.

We are anxious, of course, to make our measurements as exact as possible in most cases. Further, we tend to accept as a "ratio scale" some kinds of measures that should be interpreted only with caution. Take the problem of comparing library budgets.

If a university library has a materials budget of $600,000, a sister system $400,000, and another $200,000, we can state exact dollar and percentage differences with no problem. The larger library is 50 percent richer than the middle library and 300 percent, or three times, richer than the smallest of these three.

But what does it mean to say that one library has three times as much as another to spend on materials? Budgets, like personal income, are comparable in terms of dollars, but they risk failing to measure the meaning of what the dollars will do for the library. It may be that it takes $200,000 for any university library (compared with others in equal terms of general, educational programs) simply to maintain currency in U.S. publications. An additional $200,000 allows the library to develop collections of not only greater size but far different qualities (it can add new serial publications). And an additional $200,000 may bring the resources into still another class (it may add foreign-language materials). In effect, the character or quality of service and collections changes with such large differences in operating budget.

Likewise, in comparing the incomes of librarians or the library public. If it takes a base of $8,000 a year to maintain a family of three at some minimal threshold or standard of living, increments are not simple multiples but can (and should) be interpreted as potentially introducing altogether different life-chances that may be related to library use. Realizing this doesn't prevent us from making our analysis in terms of percentages or other ratio-scale functions, but it tells us to consider the implications in interpreting the data.

This constitutes a basic introduction to the method of assigning numbers to characteristics or values of variables which are of interest to us. We turn, in the remainder of this chapter to the computation and interpretation of some selected statistics which describe the data at hand.

We have stated that descriptive statistics serve as summary measures which help us condense large amounts of information about individual cases or subjects into a few meaningful and interpretable numbers. We will start with some very familiar measures which all have come in contact with at some time, namely, percentages, rates, and ratios. Next we will discuss some new measures which define averages for a group, give information on how similar our individual cases are, and indicate the placement of individual scores in reference to the group as a whole.

Summarizing Statistics

Some of the most useful ways of summarizing data are so commonplace that they are often overlooked. Among these are proportions, percentages, and ratios and rates. They are very useful because they allow us to compare groups of different sizes by standardizing these groups for size.

Proportion

A question may be raised about the relative emphasis given to non-English reference books in a library. Studies are made of the reference collections of the Humanities and Social Sciences divisions of a university library. The Humanities division is found to comprise 2,360 titles, of which 880 are in languages other than English. The Social Sciences collection comprises 1,890 titles, 440 of them non-English.

The proportion of non-English to English titles is expressed as 880 / 2,360 (for Humanities) and 440 / 1,890 (for Social Sciences). Simple division, indicated by the slash sign, will facilitate our comparing information about the two collections. The proportions obtained are .37 and .23 respectively. (In each collection, 1.0 would be a measure of all the titles.)

In another instance we find that, among 1,000 undergraduate library users, 250 are enrolled in independent study programs and 750 are not. The population of independent study users is 250 / 1,000, or .25. The proportion of users who are not involved in independent study is 750 / 1,000, or .75.

As a final example, of 150 fiction volumes circulated, 100 are adult and 50 are juvenile titles. The proportion of circulated fiction classified as adult is 100 / 150, or .67; the proportion classified as juvenile is 50 / 150, or .33.

Percentage

Percentages are obtained simply by multiplying proportions by 100. In the first example, cited above, we can say that 37 percent of reference titles in the Humanities division are non-English and that Social Sciences has 23 percent non-English titles.

In the second example, 25 percent of users are involved in independent study, 75 percent are not.

In the third example, 67 percent of the fiction circulation is adult, 33 percent is juvenile.

Percentages are used more often than proportions, of course, but both are common enough to remind us of the need to remember their similarity (and difference!).

In table 2 the private collection has 33 percent more fiction titles than the city collection, yet the city has 1,472 more fiction titles in

Table 2. **Data on Library Collections**

City Library Collection		Number	Percent	Private Collection		Number	Percent
Fiction		1,500	60	Fiction		28	93
Nonfiction		1,000	40	Nonfiction		2	7
	Total	2,500	100		Total	30	100

absolute numbers. Reporting percentages or proportions without sample size can be misleading to the reader.

The number of cases or observations should also be reported in our examples, the number of titles in each reference collection, the number of users, and the total circulation. This allows us to know the value, in exact numbers, of the percentage or proportion distribution.

Ratios and Rates

Ratios may be expressed in different ways. On a given day, we may have checked out 150 fiction titles and 300 nonfiction titles. The ratio of fiction to nonfiction may be expressed as 150 / 300, or as 150 : 300, or as 150 to 300. The main term is the word *to*. Whatever number comes before the word *to* is divided by the following term. Ordinarily, we like to simplify things by reducing the terms in an expression such as this. The result could be 15 / 30 or 5 / 10, or 15 : 30 or 5 : 10.

Very often we want to compare ratios and carry the reduction further, so that the second term is expressed as *unity*. In this example the ratio of fiction to nonfiction is .5 : 1, or .5 to 1. (A number less than 1 in a ratio may seem too awkward.) The same relationship of fiction to nonfiction may be expressed, alternatively, as 1 : 2.

Obviously, it could be easy to erroneously express proportional relationships as ratios, and vice versa. In the example above, for instance, the total circulation is 450. The proportion of fiction is 150 / 450, or one-third. The proportion of nonfiction is 300 / 450, or two-thirds. The ratio of fiction to nonfiction is, as indicated above, 1 : 2.

In another example, 100 adult and 50 juvenile volumes are charged out. The ratio of adult to juvenile volumes in circulation is 100 : 50, or 2 : 1.

The idea of *rate* is closely related to that of ratio, and is often used to avoid small decimal numbers. Actually, when dealing with large numbers, bases of 100 or 1,000 may be more useful. The rate of books lost in circulation might be expressed as the number of books lost per 1,000 or 10,000 circulation.

Because so many of our measures are concerned with changing conditions over time, the rate of change comes into use. A rate of change is another form of the ratio and is calculated simply by finding the difference between a variable's value at the beginning of a given period and its value at the end of the period, and dividing that difference by the value of the variable at the beginning of the period.

If you had 50,000 volumes in 1965 and 150,000 in 1975, the rate of change would be the difference (100,000) divided by the holdings at the beginning of the 10-year period (50,000):

$$\frac{150,000 - 50,000}{50,000} = 2 = 200\%$$

Your collection increased by 200 percent from 1965 to 1975.

The rate of change may be a minus value, of course, if the size of the collection has decreased.

Some people might prefer to describe the situation differently, saying that the collection in 1975 is 3 times its size in 1965. This cannot be legitimately converted into the statement that there has been a 300 percent rate of increase. Beware of reports that, innocently or not, perform one kind of operation on numbers and come up with misleading figures. This conventional procedure for rate of change may seem more "conservative," but it does precisely what it claims to do: honestly report the rate of change.

As a further example, let us say that you have been promoted to assistant department head at a salary of $15,000. Your former salary was $12,000. The percent change in your salary is calculated as follows:

$$\frac{\$15,000 - \$12,000}{\$12,000} = \frac{\$3,000}{\$12,000} = .25 = 25\% \text{ increase}$$

Summarizing Measures

Up to this point we have dealt with relatively simple sets of data, requiring little if any reorganizing. To understand two of the most important kinds of summarizing measures, we must understand a few underlying notions, particularly the *frequency distribution*.

A *frequency* is the number of times a given value occurs. If 5 students in a class score 85 on an exam, the frequency of the score 85 is 5. If 322 branch library patrons are between 20 and 29 years old, the frequency of the age group 20–29 years is 322.·

A table listing the values of a variable and the frequency of each variable value for the cases at hand is called a *frequency distribution*.

For example, the data in table 3 have been collected on the number of scheduled stops made by each of 6 bookmobiles on different routes. The variable of interest for each case (bookmobile) is "no. of stops." This information is displayed in table 4 in a frequency distribution which tells us the number of bookmobiles which made 6 stops, the number which made 9 stops, etc.

Let's assume that we join forces with 16 adjacent counties in providing bookmobile service. Our variable of interest would still be "no. of stops," but now our total number of cases under consideration would be 17. The collected data could be tabulated as in table 5, with each

Table 3. **Scheduled Stops of Bookmobiles**

Case (Bookmobile)	Value of the Variable (No. of Stops)
A	6
B	9
C	10
D	14
E	16
F	17

case (bookmobile) next to its corresponding value of the variable (number of stops) in the table.

The larger the number of cases, however, the larger and more cumbersome a table like this becomes. We need a more compact way of displaying the data, one which collapses the table into subgroups according to the value of the variable "no. of stops." If we want to summarize the pattern of stops for the total 17 bookmobiles in the area, we could use a frequency distribution such as the one in table 6. This

Table 4. **Frequency Distribution of Scheduled Stops**

Value of the Variable (No. of Stops)	Frequency (No. of Bookmobiles)
6	1
9	1
10	1
14	1
16	1
17	1

distribution presents the same information as table 5; however, it is presented in a more compact fashion. Cases are now grouped according to the value they received on the variable of interest.

The *frequency polygon* is another method by which a large amount of information can be conveyed in a concise manner for one group measured on one variable. The information in a frequency distribution is displayed graphically in a frequency polygon (see figure 1).

First, two connecting axes are drawn, the horizontal axis indicating the categories or values of the variable and the vertical axis indicating the frequency of cases at each value. Starting at a given value of the

Table 5. **Scheduled Stops for 17 Counties**

Case (Bookmobile)	Value of the Variable (No. of Stops)	Case (Bookmobile)	Value of the Variable (No. of Stops)
A	6	J	10
B	9	K	9
C	10	L	14
D	14	M	14
E	16	N	16
F	17	O	9
G	14	P	17
H	16	Q	16
I	14		

variable, we move up vertically and place a dot opposite the frequency of cases at that value. (For example, for the value "6 stops" we move along the horizontal axis until we come to the number 6; then we move up vertically and place a dot next to the 1 on the vertical axis, since only one bookmobile made 6 stops.) Each row of the frequency dis-

Table 6. **Frequency Distribution of Stops for 17 Counties**

Variable (No. of Stops)	Frequency (No. of Bookmobiles)
6	1
9	3
10	2
14	5
16	4
17	2

tribution is recorded on the graph. After all the frequencies from the frequency distribution have been recorded on the frequency polygon, the dots are connected with straight-line segments.

The frequency polygon tells us at a glance that the most frequently occurring value was 14 stops (5 bookmobiles made 14 stops). It also tells us at a glance that the smallest number of stops was 6 and the largest number of stops was 17.

The frequency polygon gives us a graphical display of the contents of a frequency distribution; it allows us to see a picture of the total data and, therefore, gives us a feeling for the group as a whole.

Another common way of graphically presenting data is the *histogram*, also known as a *bar chart*. The frequency of the variable is represented on the vertical axis. The frequency may also be written in the rectangles that represent the event. (Figure 2 is an example.)

Whether it is presented graphically or not, a frequency distribution is often too large to permit easy visual grasp of it in overall terms. Two major kinds of summarizing measures are used to describe distributions: *measures of central tendency* and *measures of dispersion*. More simply, there are averages and variations. Both kinds of measures are useful for summarizing the characteristics of a large body of data.

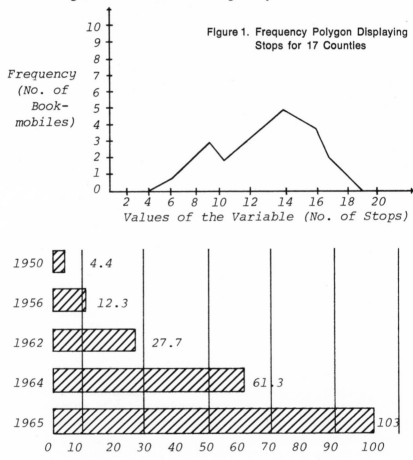

Figure 2. Histogram: Capital Outlays of Public Libraries (from all sources, in millions of dollars). Source: *Libraries at Large*, edited by Douglas M. Knight and E. S. Nourse (New York: Bowker, 1969), p. 181.

Central Tendency

The three measures we will consider here are the mean, the median, and the mode. All are known as averages; indeed each, in its own way, indicates an average.

The *mean* is the measure we most often think of as an average. The mean is computed by summing the observed values of a variable and then dividing that sum by the number of observations. The formula for the mean is

$$\overline{X} = \frac{\Sigma X}{n}$$

The symbols are

Σ = sum of
X = the value for each case or observation
n = total number of cases or observations
$\overline{X}$ = mean

$$\overline{X} = \frac{2 + 4 + 6 + 8 + 10}{5} = \frac{30}{5} = 6$$

That is, if we have the numbers 2, 4, 6, 8, 10, we arrive at the mean by adding the numbers and dividing by 5, the number of cases. The result is a mean of 6, indicated by the symbol for the mean, $\overline{X}$.

For further illustration, let us calculate the average holdings of 7 departmental libraries:

Case (Library)	Value of Variable (Volumes)
A	6,200
B	4,300
C	3,800
D	7,400
E	5,100
F	6,500
G	5,900

$$\overline{X} = \frac{6,200 + 4,300 + 3,800 + 7,400 + 5,100 + 6,500 + 5,900}{7} =$$

$$= \frac{39,200}{7} = 5,600$$

Therefore, 5,600 is the mean number of volumes for the 7 libraries.

The *median* is that value which is midway in a frequency distribution in which all values are ranked from highest to lowest. There is, then, an equal number of cases above and below the median. If we have the numbers 2, 4, 6, 8, 10, the median is 6. The symbol for the median is $\dot{X}$.

In the example of the 7 departmental libraries, the median is obtained by rank-ordering the holdings from lowest to highest, or vice versa, and selecting the middle case.

7,400
6,500
6,200
5,900
5,100
4,300
3,800

The median of the departmental libraries is 5,900. Three libraries have more volumes and 3 have fewer volumes.

If we take the numbers 2, 4, 6, 8, 20, the median is still 6. But what about the mean? It now is 8. The median is less affected by extreme values than is the mean. The mean, by definition, gives "equal opportunity" to all values; they are represented in direct proportion to their weight. In a strict mathematical sense, it is the truest, most representative measure of central tendency because it takes all values fully into account. However, there are times when summarizing data by the mean may seem to be misleading, as in the last example.

Salaries and income are among those variables that sometimes tend to be skewed upward. Consider the following cases.

Group A		Group B	
Case	Income	Case	Income
1	$ 2,000	6	$ 2,000
2	4,000	7	4,000
3	6,000	8	6,000
4	8,000	9	8,000
5	10,000	10	22,000

In group A, both the mean and median are $6,000. In group B, the median is $6,000 but the mean is $8,400. As four out of five people earn less than the mean in group B, the median is a more representative measure. Salary distributions are often similar to those in group B and, consequently, the median is often used to report salaries.

Both the mean and the median are mathematically correct, of course. Rather than worry about which of the two is a better summary, it makes more sense to calculate each one and report both, especially when there is a substantial difference between them. The reader is thereby provided more information, which will allow for better appreciation of the significance of the salary distribution.

When there is an equal number of cases, locating the median requires a minor computation. In the array 2, 4, 6, 8, 10, 12, there is no stated value above and below which there is an equal number of cases. To compute the median, add the two central cases, 6 and 8, and divide by 2 to arrive at the median, 7.

The *mode* is that value of the variable which occurs most frequently. In the samples given so far, there was no mode; each value appeared only once. We might have 9 library staff members earning as follows: $4,100, $6,000, $6,000, $6,000, $8,000, $9,000, $10,000, $11,000, $20,000. The mode is $6,000. The symbol for the mode is $\hat{X}$.

What are the values for the mean and the median in this distribution? Figure 3 illustrates their placement in relation to each other.

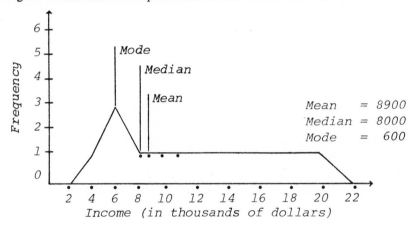

Figure 3. Frequency Polygon Displaying Staff Salaries for 9 Librarians, Unimodal Distribution

The mean is calculated by adding all the salaries (as follows) and dividing by the total number of salaries:

$$
\begin{array}{r}
\$\ 4{,}100 \\
6{,}000 \\
6{,}000 \\
6{,}000 \\
8{,}000 \\
9{,}000 \\
10{,}000 \\
11{,}000 \\
\underline{20{,}000} \\
\$80{,}100
\end{array}
\qquad
\bar{X} = \frac{\$80{,}100}{9} = \$8{,}900
$$

The median is $8,000—above and below which there are equal numbers of cases.

A distribution may be multimodal, and bimodal distributions are especially common. On a large library staff, a large number of people may earn $6,000 and another large number of people $9,000, as displayed in figure 4. In such a case, one might suspect that two different groups

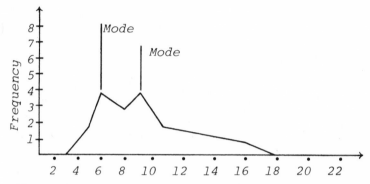

Figure 4. Frequency Polygon Displaying Staff Salaries, Bimodal Distribution

were lumped together as "staff." The modes suggest one average or central-tendency salary for subprofessionals and another for young (or at least beginning!) professionals.

Measures of Dispersion

Measures of dispersion enable us to see variations from the central tendency of data. Consequently, we often use, or find reported, one or more of several standard measures that add to our understanding of the summarizing quality of a central tendency.

The *range* of data in a frequency distribution tells us the difference between the highest and the lowest quantities. It is that amount between the lower and upper limits of the data. If the lowest beginning professional salary at any given year is $8,000 and the highest is $12,500, the range is $4,500 ($12,500 — $8,000). The range is generally expressed in terms of its lower and higher limits: $8,000–$12,500.

The range can be used to compare variabilities. Beginning salaries in the Northeast are $6,700–$9,200 and in the Southeast $8,000–$9,400. The ranges are, respectively, $2,500 and $1,400.

The *mean deviation*, although no longer so common, is still a useful way to describe variation about the mean. It tells us the average (mean) amount of variation around the mean $(\overline{X})$ for a group of values. Each value is represented by an X. The difference from the mean, $X - \overline{X}$, is expressed by x. The formula for the *mean absolute deviation* is $\frac{\Sigma|X - \overline{X}|}{n}$.

Take the sum of the absolute values of the difference between each observed value and the mean, and divide that sum by the number of observations. The vertical bars enclosing $X - \overline{X}$ stand for "take the absolute value"; that is, make any negative difference positive and leave the positive differences positive.

For the data in table 7 the mean weekly circulation is 1,600. The mean deviation is obtained by calculating each week's difference from

Table 7. **Apex Branch Circulation Figures for Each of Past 6 Weeks (Listed in X Column)**

X	$X - \overline{X}$ (or x)	Absolute Value of x	
1,200	-400	400	
1,900	300	300	$\overline{X} = \dfrac{\Sigma X}{n} = \dfrac{9,600}{6} = 1,600$
1,600	000	000	
1,100	-500	500	
2,000	400	400	$MD = \dfrac{\Sigma x}{n} = \dfrac{1,800}{6} = 300$
1,800	200	200	
$\Sigma X = 9,600$		$\Sigma x = 1,800$	

the mean, adding the *absolute value* of these differences, and dividing by the number of weeks.

A summary statement of these data can now be expressed: $\overline{X} = 1,600$, MD = 300. These figures tell us that for the 6-week period the average weekly circulation was 1,600 volumes and the average variation around this mean was 300 volumes. On the average, then, 1,600 volumes per week were circulated and the weekly variation averaged 300 volumes, greater or less than the average for the 6-week period.

Standard deviation is the most important measure of dispersion. While it is somewhat more complicated and a bit more difficult to calculate than the range or mean deviation, it is very useful. The standard deviation is generally calculated because it has convenient mathematical properties which statisticians prefer. It is reported in the literature more frequently and is fundamental to working with inferential statistical procedures.

To calculate it, take the deviation of each score from the mean, square each of these differences, sum the results, divide by the number of cases, and take the square root. The squaring of the variations from the mean is a standard mathematical practice.

In the average deviation, we are asked to ignore the signs, but this seems somewhat uncomfortable mathematically. If we square the variations, we get rid of the signs and end with an absolute difference from the mean, regardless of the sign. Furthermore, squaring the variations gives the outer limits some extra weight and gives us some basis for estimating the greater range of variation we might find in the total population. We take the square root to bring us back to the immediate reality of our sample.

The general formula is

$$s = \sqrt{\frac{\Sigma (X - \overline{X})^2}{n}}$$

If we use the data in the last example, the standard deviation can be computed as follows:

Weekly Circulation	Variation from 6-Week Mean	Variation Squared
X	$(X - \overline{X})$	$(X - \overline{X})^2$
1,200	−400	160,000
1,900	300	90,000
1,600	000	0,000
1,100	−500	250,000
2,000	400	160,000
1,800	200	40,000
		700,000

$n = 6$
$\overline{X} = 1,600$
$\Sigma (X - \overline{X})^2 = 700,000$

$$s = \sqrt{\frac{700,000}{6}} = \sqrt{116,666.67} = 341.57$$

The interpretation of the standard deviation is simple if the shape of the distribution is normal. (A discussion of the normal distribution appears a few pages further on.) For the moment, suffice it to say that the mean "plus-and-minus 1" standard deviation includes two-thirds of the cases in a normal or near-normal distribution. In our example, 1,600 ± 342 includes 4 out of 6, or two-thirds, of the cases.

Consider the following (from table 8), which depicts the number of professional personnel employed in various types of libraries in the Southeast. If we focus on the public library, we can summarize as follows:

Case	No. of Professional Personnel	
Alabama	119	
Florida	205	
Georgia	238	
Kentucky	63	$\overline{X} = \dfrac{1,387}{9} = 154$ (rounded to
Mississippi	112	nearest whole
North Carolina	127	number)
South Carolina	85	
Tennessee	177	
Virginia	261	
Total	1,387	

Calculating the standard deviation produces a figure of 65.

$$\overline{X} \pm s = 154 \pm 65 = 89 - 219$$

We expect to and *do* find that the standard deviation around the mean includes 5 of the 9 states, a nearly normal distribution.

Because it is a standardized measure of dispersion, the standard deviation may be used to compare the equality or inequality of two or more groups. If groups are comparable, the larger the difference in standard deviations, the greater the inequality. Frequently the comparison of standard deviations alone can lead to incorrect interpretations. One method of ensuring against this is the *coefficient of variation*.

Coefficient of Variation

Sometimes standard deviations differ a great deal but do not necessarily reflect the true difference in homogeneity of inequality. For instance, a standard deviation of a municipal library's branch budgets of $1,000 in 1970 and $1,300 in 1975 doesn't mean that the economic inequality among branches necessarily grew worse in that 5-year period.

A comparison of the two periods should be based on the ratio of the standard deviation to the mean: the coefficient of variation. Knowing the mean budgets for 1970 and 1975, we can easily calculate this coefficient: $V = s / \overline{X}$.

Table 8. Number of Professional Personnel by Type of Library in Which Employed, by State and Region

State	Academic Library	Library Education	Public Library	School Library Media Program	Special Library	State Library*	Total
Alabama	125	17	119	151	30	10	452
Florida	305	10	205	424	28	16	988
Georgia	277	21	238	952	27	17	1,532
Kentucky	102	11	63	668	18	14	876
Mississippi	118	10	112	118	13	4	375
North Carolina	307	24	127	395	30	20	903
South Carolina	140	8	85	225	29	15	502
Tennessee	199	15	177	260	38	7	696
Virginia	341	3	261	742	44	41	1,432
Region	1,914	119	1,387	3,935	257	144	7,756

*Includes state library agencies, state libraries, and state school-library supervisors. Source: Mary Edna Anders, *Survey Tables* (Atlanta: Georgia Institute of Technology, 1975), p. 8.

1970: Mean branch library budget	$10,000
Standard deviation	1,000

$$V = 1,000 / 10,000 = .100$$

1975: Mean branch library budget	$14,000
Standard deviation	1,300

$$V = 1,300 / 14,000 = .093$$

The inequality among branch library budgets has actually decreased from 1970 to 1975. While the standard deviation became larger, the average budget increased relatively more. Precisely, the rate of change is $(.100 - .093) / .100 = .07$, or a 7% decrease in inequality.

The Normal Distribution

Statistics is a branch of mathematics; therefore, we have to learn about certain families of mathematical distributions to be able to use and understand statistical methods.

The first distribution which we will consider is the family of *normal distributions* or *normal curves*. Figure 5 is a normal distribution, taken from the family. All normal distributions have the same characteristics. The only way in which they may differ is in the *mean and/or standard deviation* of the distribution. The points which make up the curve, as in figure 5, are derived through a special formula (which will not be given here), and each member of the family of normal curves can be found by means of this formula. The only element(s) which differ in this formula from curve to curve are the mean and/or standard deviation.

All the distributions which we will deal with in this text have a corresponding formula from which the points on the curve can be derived. These distributions are pictorially similar to frequency distributions. They are different in that they are "smooth" or "continuous," having an infinite number of points on the curve. Also, as we have stated before, they are derived from formulae, whereas our frequency distributions are created from empirical, or observed, data. Where we had the value of the variable plotted against its frequency in our frequency polygon, we have the value of a variable plotted against its likelihood in our mathematical probability distribution.

The family of normal distributions has characteristics that are useful in conjunction with the standard deviation, which will be outlined in the section on inductive statistics. These characteristics are:

1. The mean, median, and mode of a distribution have the same value. If we were to draw a vertical line through the curve, so that the area under the curve was divided into two equal portions, the line would intersect our horizontal axis at the value which represents these three measures of central tendency for the distribution.

2. The distribution is symmetrical. Areas on each side of the bisecting dotted line below are equal. This means that an equal number of cases is distributed on each side.

3. The ends or tails of the curve extend infinitely. They do not touch the base line, and are said to be asymptotic.

4. There is an infinite or indefinitely large number of cases under the curve (see figure 5).

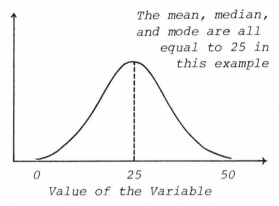

The mean, median, and mode are all equal to 25 in this example

Value of the Variable

Figure 5. A Normal Distribution with Population Mean Equal to 25

Because many events in life are nearly normal in distribution, this special curve is commonly used. One sense in which it may be said to be normal is that most of the cases fall near the center of the curve. Further, if we draw perpendicular lines from the base at 1 standard deviation on either side of the mean, the area under the curve between the mean and each of these lines will include 34 percent of the cases on each side of the mean. Thus the notion of "standard"—1 standard deviation unit from the mean—includes 34 percent of the cases in a normal or near-normal distribution, regardless of the size of the distribution.

In figure 6, 1 standard deviation unit is marked off on either side of the mean. The unit to the left of the mean will include 34 percent of the cases of less value than the mean; the unit to the right will include 34 percent of the cases of greater value than the mean. Collectively, they include 68 percent, or just over two-thirds, of the cases.

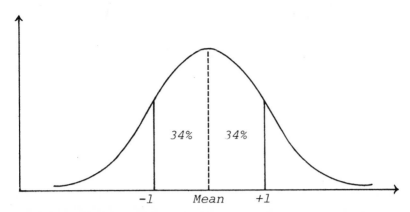

Figure 6. A Normal Distribution, Indicating 1 Standard Deviation Unit above and below the Mean

Because the deviation unit—the distance on the base line from the mean—is standard, it is possible to move equal distances on this base line and obtain fixed proportions under the curve for each unit. In figure 7, note that, in doing so, increasingly smaller proportions are added because of the slope of the curve. Approximately 13 percent more is added by the second standard deviation, and approximately another 2 percent by the third standard deviation.

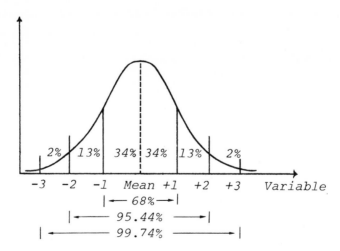

Figure 7. A Normal Distribution, Indicating Placement of 1, 2, and 3 Standard Deviation Units on Either Side of the Mean

More often than not, our interest is in the value of 1 or more standard deviations on both sides of the mean. The mean plus-and-minus-1 standard deviation then includes 68 percent of the cases (34% + 34%), which is often reported as "roughly two-thirds"; the mean plus-and-minus-2 standard deviation accounts for 95 percent of the cases; and 3 standard deviations on both sides of the mean incorporate just under 100 percent of the total number of cases.

The standard deviation assumes major importance in inductive statistics. When it is used simply to summarize a total set of data, it describes the dispersion in its own particular terms. Indeed, the mean absolute deviation may often be more useful for purely descriptive purposes. The standard deviation will also tell something about the departure of the data from a normal distribution—if, for instance, it doesn't truly account for approximately 68 percent of the cases around the mean. Nonetheless, the standard deviation is a convention which may be used for descriptive analysis when no inferences are clearly intended, if only because it is fairly easy to compute and has a general, intuitive utility and appeal.

Z Scores

Z or *standard scores* are directly derived from the standard deviation. They can be used with any type of distribution, but are most useful when the shape of the distribution is nearly normal; that is, when the mean and the median are very similar and taking 1 standard deviation unit on each side of the mean includes two-thirds of the cases. A Z score is expressed in terms of standard deviation units and indicates how many standard deviation units above or below the mean a score (or case) is.

Many people have had experience with being "located" in Z-score terms, perhaps unwittingly! Intelligence tests have been applied so widely as to enable us to use the distribution of IQ scores as a standard normal distribution. For instance, IQ scores have a mean of 100, with a standard deviation of 15; thus 68 percent of the population have scores between 85 and 115. Someone with an IQ score of 130 is 2 standard deviation units higher than the mean and thus is (approximately) among the highest-scoring 2 percent of the population.

The following example will demonstrate how this principle may be applied in a library situation.

Z Score Sample Problem

A survey of library administrators' salaries reports a mean salary of $17,940 and a standard deviation of $4,960, and an administrator who earns $25,000 is interested in knowing where he or she falls in the distribution. He or she believes it is reasonable to assume that the administrators' salaries are normally distributed. To calculate the relative position, we may use the formula

$$Z = \frac{X - \overline{X}}{s} = \frac{\$25,000 - \$17,940}{\$4,960} = \frac{\$7,060}{\$4,960} = 1.42$$

The 1.42 is the number of standard deviation units away from the mean that the $25,000 salary falls. Because it is higher than the mean, we know that it will fall that many units above the lower 50 percent. By using the table "Areas under the Normal Curve" (appendix table 2), we can locate the percentage value of the 1.42 standard deviation units.

Reading down the outside column to 1.4 and over to the figure under the column .02, we find the figure .4222. Converted to percentage, this is 42.22 percent, or 42 percent rounded. This means that the salary is greater than 42 percent of the salaries that are greater than the mean salary, or, in other words, the salary is greater than 92 percent of all the salaries (since the mean salary is greater than the lower 50 percent of all the salaries.) Thus 92 percent are paid less and 8 percent are paid as much as or more than the administrator who earns $25,000. (Notice that this calculation is based on the assumption that library administrators' salaries are nearly normally distributed.)

Chebyshev's Theorem

According to Chebyshev's theorem, the standard deviation can be used to determine the areas under a curve that are not normal. We sometimes have reason to believe that the distribution we are working with is multimodal and that it is not reasonable to apply the standard deviation as if the curve were normal. We may have a distribution that looks like figure 8. With such a curve, we may apply the following formula:

$$100 \left(1 - \frac{1}{Z^2}\right)$$

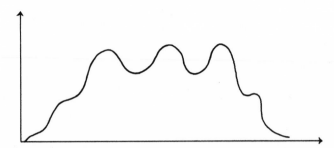

Figure 8. A Multimodal Distribution

If, in our example of administrators' salaries, we know the distribution of these salaries to be multimodal, we can use the above formula to modify the normal Z interpretation. Substituting in the formula, we obtain these results:

$$100 \left(1 - \frac{1}{1.42^2}\right) = 100 \left(1 - \frac{1}{2.02}\right)$$
$$= 100 \left(1 - .495\right)$$
$$= 100 \left(.51\right) = 51\%$$

We obtain a much more conservative estimate of the relative location of our example salary. Rather than in the top 8 percent (given a normal distribution) we now assume that our administrator is in the top 49 percent of his peers' salary distribution.

Exercises

1. Refer to table 8 (Number of Professional Personnel by Type of Library in Which Employed, by State and Region).
a. What percent of North Carolina's librarians are in public libraries?
b. What are the mean, range, and standard deviation of the distribution of public librarians in the Southeast?

2. Refer to table 6, which contains a frequency distribution of number of stops for seventeen counties.
a. Calculate the mean, median, and mode of the distribution. (Remember that there are 17 cases.)
b. Determine the standard deviation.
c. Find the Z score which corresponds to a value of 16 on the variable.
d. Construct a histogram of the data.

REFERENCES AND SUGGESTED READINGS

Babbie, Earl R. *The Practice of Social Research*. Belmont, Calif.: Wadsworth, 1975.

Blalock, Hubert M., Jr. *Social Statistics*. 2d ed. New York: McGraw-Hill, 1972, p. 11–88.

Fields, Craig. *About Computers*. Cambridge, Mass.: Winthrop, 1973.

Glass, Gene, and Julian C. Stanley. *Statistical Methods for Education and Psychology*. Englewood Cliffs, N.J.: Prentice-Hall, 1970.

Hoadley, Irene, and Helen Clarke, eds. *Quantitative Methods in Librarianship: Standards, Research, Management*. Westport, Conn.: Greenwood, 1972.

Huff, Darrell. *How to Lie with Statistics*. New York: Norton, 1954.

Kerlinger, Fred N. *Foundations of Behavioral Research*. 2d ed. New York: Holt, Rinehart and Winston, 1973.

Loether, Herman, and Donald McTavish. *Descriptive Statistics for Sociologists*. Boston: Allyn and Bacon, 1974.

Sanders, Virginia. *Measurement and Statistics*. New York: Oxford University Press, 1958.

Siegel, Sidney. *Nonparametric Statistics for the Behavioral Sciences*. New York: McGraw-Hill, 1956.

2
Sampling

In chapter 1 we dealt with descriptive statistics, which is a method by which we summarize characteristics of a group of cases on specific variables into a few interpretable numbers (i.e., mean and/or standard deviation). The term descriptive statistics arose because we describe the group at hand by means of the data we have gathered. At some point we may want to make inferences about a population of cases from the information we gather on a small subgroup of that population, which is called a *sample*, and the process of selecting that sample is guided by the rules of sampling theory. All formal statistical tests are based upon the assumption that the sample under consideration is random. Therefore, sampling and sampling theory play an extremely important part in inferential statistics. The concept of randomness, along with other sampling terminology, will be presented in this chapter.

We hope to provide an understanding of the major considerations in deciding on sampling plans. The technology of sampling has become so highly developed that a major research effort, involving a good deal of sampling, requires the advice of a sampling expert. It is important, however, that librarians, like other researchers, provide the substantive information essential for drawing up samples with the aid of sampling experts. Moreover, in many instances the individual library researcher or the research team can develop an adequate sampling plan without technical assistance. The following should provide a sound basis for understanding the basic problems and general guidelines in sampling procedures. Table 9 contains an outline of the types of samples which will be discussed in this chapter.

The purpose of sampling is to provide information that will allow us to make generalizations about the population being sampled. A *population* is any aggregate of persons or things that have traits or characteristics which we want to measure. If we are interested in library users, the population consists of library users. The term universe is often used for the term population.

Table 9. **Types of Sampling Techniques under Consideration**

Probability	Nonprobability
Simple random*	Purposive
Systematic*	Quota
Stratified*	Accidental
Proportional	
Disproportional	

*Indicates that data may be used in formal inferential techniques.

In some instances, sampling from a population is not necessary because there may be greater virtues in taking a complete census. For example, when a researcher wants to analyze a population that is relatively small, a complete census may be more appropriate than a sample. A college library staff, serving a total undergraduate body of 900 students, or a branch library, serving a similarly small number of patrons, may be well advised to study all their users. The cost would not be great and the reliability of the results would be more certain. Similarly, if one is interested in finding the average cost of new books on order and if the number of books is only 100, it would be simple to calculate the average cost and avoid sampling.

A complete census will often provide more acceptance by the public in terms of the findings and the responses to inquiries for data. There may be greater validity in findings that are based on a total census, of course, and it may be easier to enhance the validity of one's findings or data gathering. In addition, taking a complete census precludes the need to consult a sampling expert.

In practical terms, however, we often do not have the time or money to analyze the population that we are studying, and we want to do the best we can with limited resources to arrive at some conclusions about this totality. In such situations, sampling has a clear advantage. As far as validity of the results is concerned, a sample, if correctly taken, may do as well as a complete census. Furthermore, not even unlimited amounts of time or money will necessarily be adequate to obtain the skilled personnel and other services required for an extremely large sample, let alone a total census.

Probability Sampling

Samples fall into two major classes, *probability* and *nonprobability*. The theory of statistical hypothesis testing is based on the assumption that the sample under examination is one in which each case or element has a known probability of occurrence; that is, it is a probability sample. Probability sampling must be done by some method of random sampling and probabilities of occurrence must be taken into account when making estimates from the sample. The standard errors and biases of various sampling plans have been studied in terms of probability theory, and our knowledge of them allows us to choose the most suitable plan for our needs. Probability sampling also allows us to compute from our sampled data the standard error of our estimate and the confidence limits (discussed later) for the true population value, giving us an idea about the accuracy of our estimate. If we use nonprobability sampling, we really have no way of evaluating the accuracy of our estimate.

Our purpose in sampling is to measure certain characteristics of the elements in a sample in order to estimate the nature of these characteristics in the population. A sample must be as representative of the population as possible in order to make valid estimates, and the best strategy for ensuring representativeness is probability sampling.

Simple Random Sampling

The basic type of probability sampling is the *simple random sample*. In simple random sampling, each member of a population has an equal chance of being included in the sample; the sampler does not exercise his or her own judgment in picking members or in excluding them from the sample. Although many discussions of random sampling are highly mathematical and mystifying, the fact is that random samples are easily understood, require minimal mathematical reasoning, and are easily drawn, thanks to "prepackaged" labor in the form of tables of random numbers.

Let us assume that you want to draw a sample of 80 graduate students as users of a Science and Technology division library and that there is a list of the names of the 480 students enrolled in the departments in this division. This group of 480 constitutes the population to be sampled. You may simply number each name once, beginning with 1 and ending with 480; turn to any table of random numbers (see appendix table 1); select any 3-digit row (or any number of rows that will provide 3 digits); and begin at a random point. Do not deliberately select as a starting point a number that will include some favorite case or otherwise influence the selection. Simply read the 3-digit numbers down or across and extract all numbers from 1 through 480 until you have a total of 80 cases. Numbers larger than 480 are passed over (you have no name with a number higher than 480). Any number that is repeated is passed over after its first appearance and selection. The procedure is so simple that it may seem unbelievable or even confusing.

As a trial, pretend that we are to select a sample of 5 from a population of 50. After we turn to the table of random numbers in the appendix of this book, table 1 (any table of random numbers may be used, of course), we begin our random start in the sixth row and the fifth column, where we locate the number 72682. This table consists of 5-digit numbers, but we need only the first two digits of each number. We pass over the first number, 72, as we have no element numbered above 50. Reading downward, we find that we can use 21 and then 01. Number 80 is too large, but 13 can be selected. Number 21 appears next, but it duplicates a case we've already selected; so we pass it by. Our final choices are the next two numbers, 47 and 40. Thus our sample will consist of the names numbered 01, 13, 21, 40, and 47.

An actual sample will usually be much larger but the procedure is the same. All that is required is a list or inventory that is or can be numbered and a table of random numbers. (There are computer programs with random-number-selection capability—saving one even this simple chore—but only large samples usually warrant a computer selection.) In drawing a sample manually, you may use any number of adjacent rows that is necessary. When you have used all possible numbers from one row and need more numbers to complete the sample, return to the top of the page and begin with another set of adjacent rows.

Formal tables of random numbers (or their computerized counterparts) are virtually foolproof, as long as you follow the prescribed procedure. The numbers are truly random and are the basis for scientific sample selection in all fields.

Not infrequently there are problems with the list of the population to be sampled. First of all, it is mandatory that each element (name, title of book, whatever) appear on the list only once. Any element that appears more than once obviously has a biased chance of being selected. Lists may need "purification" to make certain of this and, otherwise, to provide an accurate representation of the population. Other problems are lists that may be out of date, with names omitted that need to be added or with names of persons no longer alive or those who should be omitted. Up-to-date, accurate lists are often difficult to find. Rarely do lists not need checking for errors of omission or commission.

Who hasn't heard of the election-prediction fiasco in the 1930s, when a prominent magazine decided to poll the electorate, using the telephone directory as a list? In those days, many people did not have a telephone and, consequently, were not in the list. Voters who did not have a telephone turned out in such large numbers that they elected Franklin Roosevelt, and the magazine's erroneous prediction of election for his opponent was a major cause of its demise. Telephone users may indeed have voted for Mr. Landon, but those other voters, unrepresented in the telephone list, greatly outnumbered them.

Even today, telephone directories may not include low-income people, who may be important for the research at hand. City directories are often so out of date by the time they are published that they, too, need purification. Also, lists that have duplications can be a serious problem unless the duplications can be easily identified. Membership or organization lists, staff directories, and the like often contain the same names in several places—as anyone who receives "junk" mail can testify.

The issue generally boils down to "cleaning" the list as best you can and proceeding only when you feel assured that it is adequate. Consulting people who have special knowledge about the population can be helpful. And sometimes the research problem can be redefined so that the population will "fit" the list.

Always bear in mind that a heavily biased list may be worse than no list at all. Far more likely than not, the former will lead to very erroneous inferences.

Systematic Sampling

A common and extremely useful alternative to simple random sampling is *systematic sampling*. Drawing a simple random sample from a very long list can be very tedious. (Imagine sampling from a list of 150,000, especially if the list is not already numbered!) Drawing a systematic sample, on the other hand, is very easy.

Rather than use a table of random numbers, we randomly select the first element as a starting point and proceed to select elements at a fixed interval, k. If we were to draw a sample of 40 cards from an order file of 1,000 cards, we would select every 25th card [(1,000 / 40) = 25]. We would begin by selecting our first element at random from one of the first 25 cards and each 25th card thereafter. Let us assume that our first element is the 10th card; our next card will be the 35th, then the 60th, the 85th, and so on, to the 985th card, yielding a total of 40 cards.

The first element must be selected at random. We might use a table of random numbers, or ask a colleague to give us a number from 1 to 25.

The assumption of randomness may be made as long as there is no reason to believe that there is anything in the ordering of the list to introduce nonrandom variation in the variables that are being measured. A problem can arise where the interval coincides with some characteristic in the list. There is little likelihood of such a problem in library research, but an example may point up the possibility.

A public librarian, making a survey in a neighborhood development, might find that his or her random start is a corner house and that all blocks in the neighborhood have a number of houses which correspond to the size (number) of the sampling interval. Result: a sample of corner houses whose inhabitants, for various socioeconomic reasons, may have different responses about public library use than people who do not live on a corner.

Many systematic samples in library research employ alphabetical lists such as bibliographies. Rarely should there be any reason to suspect that anything is inherent in the alphabetical order that will violate random association of the variables being measured. Indeed, lists of authors (or users) may present some clustering of ethnicity (e.g., Mc's and O's), and produce a sample that has a reasonable proportion of ethnic measurement, but rarely would such a distribution interfere with the measurement of problems in librarianship, so far as sampling is concerned.

Given the very large files which may be used as lists in library research, systematic sampling is a very handy solution. Consider taking a fairly large sample from a large shelf-list catalog of hundreds of thousands of cards. Numbering each card and drawing a sample from a table of random numbers would be extremely difficult, but systematic sampling can reduce the labor enormously. Indeed, if we can assume that there are X cards per inch, we can sample by intervals: in inches or centimeters or some other linear interval.

No matter which technique is employed, it is important that the researcher analyze the list for defects and correct them. One danger in using this "distance interval" is that older cards tend to wear down and may not be prominent enough to be selected at any given interval when they are adjoined by newer, less worn cards.

Stratified Sampling

In some instances, *stratified sampling* may improve the design of the sample. Stratifying a sample is performed by classifying or dividing all the elements in the population into categories that are relevant for the study. Each element, of course, must appear only once, and in only one stratum. In *proportional* stratified sampling, we take a simple random or a systematic sample from each stratum so that we produce a set of subsamples that are directly proportional in size to their occurrence in the population.

If, say, in surveying the 100 libraries in a region, we know that 50 percent are school libraries, 30 percent are public libraries, 15 percent are academic libraries, and 5 percent are special libraries, we have the information to stratify our total sample so that each type of library is represented in the proportion it has in the population (see figure 9). (We may, however, want to stratify *disproportionately*, and the reason for this will be clarified below.)

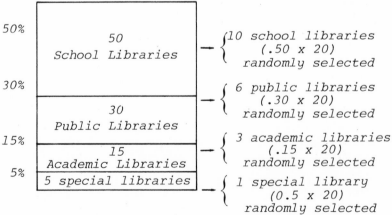

Figure 9. Illustration of Proportional Stratified Random Sampling

Stratifying can make our sampling procedure easier. We may have separate lists for each type of library and thus find it easier to sample from each list than to combine all lists and take an overall random or systematic sample. In fact, when there are several long lists it is often unfeasible to combine them for a single random sample.

Stratification may also make our sample more efficient. In general, we find that stratification improves our sampling *if* all the strata are internally homogeneous. In other words, if the variations in the *characteristics* that are being studied are more homogeneous within than between strata, we gain improvement by stratifying. If the strata are highly heterogeneous—the variations are greater within each stratum than between the strata—we gain little or nothing.

When we measure several variables, the situation is more complicated. Some variables may be similarly distributed throughout the population (across strata) while others may vary more within strata than between or among the strata. For instance, per capita budget expenditures may not vary greatly from one type of library to another, and stratifying for this purpose would gain us little. However, another variable, such as per capita circulation, may vary widely among the strata. The important thing is to make certain that there are enough elements (libraries) within each stratum (type of library) to provide a basis for assessing the variations within each stratum.

Disproportional stratified sampling can give us such assurance. In the example cited above, we had only 5 special libraries in our population list. With a simple or systematic random sample, or a proportional stratified sample of 20 percent, we would end with only 1 special library and, consequently, have too few data to determine the range of variation in special libraries in the variables being analyzed—per capita budget or per capita circulation, for instance. We may therefore want to include all 5 of the special libraries, as a subsample, to make certain that we have an accurate assessment of the *range* of behavior within that group. Indeed, we may want to "oversample" academic libraries and public libraries. Our decision to do so will depend in large part on whether we know or suspect that the variable in question is strongly related to the type of library. We might know or suspect that budgets are very much alike among school libraries and that they vary considerably among special or other types of libraries.

As in simple random sampling, the purpose of stratifying is to capture the *range of variation* in the *variables* under study. In other words, stratification is not intended, as is quota sampling, to produce a micromodel replica of the population in terms of proportional representation. The decision to stratify is based on whether the traits or characteristics under study are homogeneous within strata, compared with their occurrence between strata.

Another factor in deciding on stratification is its cost. As in all sampling procedures, the cost of the sample may be directly related to precision. Our decision to select a particular sampling procedure must take into account what we are willing to pay for a given level of precision.

Correcting Disproportionately Stratified Samples

When a sample is disproportionately stratified it is possible to summarize results within each stratum without specific corrections. However, summarizing for the entire sample, over all strata, will require a weighting procedure to correct for the disproportionate representation of strata. In fact, this weighting may be used for either stratified samples or for a type of nonprobability sample that is known as a quota sample.

As an example, let us assume that we have polled 500 adults in the community. We have asked them if they use the library and whether they are high school graduates. Our results are as shown in table 10 (by the figures *not* in parentheses).

Table 10. **Data for 500 Adults on Library Usage and Graduate Status**

Use Library	Not High School Graduate		High School Graduate		Total	
Yes:	10	(20)	200	(150)	210	(170)
No:	90		200		290	
	100	(200)	400	(300)	500	

However, let us assume we know from the U.S. Census or other reliable sources that only 60 percent of the adults in the community are high school graduates and that 40 percent did not complete high school, as contrasted with the 80 and 20 percent in our sample. Consequently, if we had a sample proportionately stratified in terms of high school graduation, the figures in the parentheses at the bottom of the rows would be expected. Further, as half of the high school graduates in our survey are known to be library users (200 of 400), the figures in parentheses in the "yes" row would be expected.

As a result, if we were to sample proportionately—or correct for known disproportions in terms of education level—we would find that 34 percent of the community adults are library users, rather than 42 percent. The initial higher percentage was produced as a result of not accounting (by stratifying) for level of education that is actually characteristic of the community. The correction for level of education would presumably yield a more accurate estimate of the *population's* library use. In effect, we have corrected a sample bias or misrepresentation, and have produced an important change in results by doing so.

Nonprobability Sampling

Purposive Sampling

Sometimes economic considerations make probability sampling difficult, and so we must select a less desirable sample, known as a *nonprobability* sample. Perhaps the most important type of nonprobability sample is the *purposive* sample, sometimes known as the *experience* sample.

In exploratory research we are often interested in obtaining information on a subject that is relatively unknown. Information of high quality may best be obtained by turning to experts or people with great experience or sensitivity to the subject, rather than to a random selection among all people involved.

For instance, investigation of the interrelationships between the public library and municipal or county officials might produce better results from data obtained from public library directors, who are known to be very active in their communities and highly sensitive to their social and political roles, than from a sample of all members of library staffs. Of course, findings from such a study cannot be used to generalize about the experiences or perceptions of all public library directors, much less all public librarians. To the extent that we are sure that our "experience" sample of experts is representative, we may try to generalize to the special "universe" of public library directors who are aggressive leaders.

We can employ two methods to test the representativeness of such a sample. One, we could use our initial findings as a basis for a questionnaire or interview and administer it to a random sample. Two, we could look for independent evidence, such as might appear in annual reports, library literature, or observation of directors' behavior. Obtaining similar responses would encourage us to think of our sample as representative.

Quota Sampling

Another type of nonprobability sampling is *quota* sampling, a procedure that attempts to produce a replica of the population in terms of certain characteristics. If it is known that 15 percent of a library's users are high school students, field workers would be directed to include that quota or percentage of such people in their observations.

Although it is sometimes so represented—and can with rather careful precautions approximate a random sample—quota sampling is not random or probability sampling. Quota sampling is widely used by pollsters and marketing researchers, with reasonable effectiveness, for many purposes. (It should be noted, however, as one indication of the undesirability of quota sampling, that the Gallup Poll decided in 1972 to abandon the quota system and adopt random sampling.)

Accidental Sampling

Quota sampling is one step removed from accidental sampling—the selection of a sample in the most convenient way possible, with no assurance that the cases will be representative of the population one wants to study. An accidental sample of users could consist of the first 50 people who come to the circulation desk. An accidental sample of library materials would be the first 50 new books that arrive in the mail. Quota sampling goes further than this and *attempts* to provide some minimal provision for certain selected characteristics.

For instance, we could improve the accidental sample of 50 users by taking the first 25 women and the first 25 men. If there are differences in use related to the sex of the user, we will have some definite basis for studying these differences. We might also apply a "quota" system to the accidental sample of 50 new books. If a variable—data of publication, for instance—were of interest, we might want a quota of materials in the Humanities, the Social Sciences, and the Sciences. We could therefore take the first 17 titles in each of these major fields as they arrive.

However, such tactics in no way provide any assurance that the quotas will yield a representative sample. After all, the primary goal of sampling, given economy and time, is to provide a group that is *representative* of the population. Our single or best assurance for representative sampling is probability sampling. Furthermore, the virtues of identifying quotas can be accommodated by a special kind of probability sampling, discussed previously as *stratified* random sampling.

Sample Size

The formula for determining the size of the sample is

$$n = \left\{ \frac{Z\,(\sigma_g)}{E} \right\}^2$$

where n = sample size

    σ_g = the guessed value of the standard deviation

    E = the amount of error we will allow for

    Z = the accuracy we estimate as desirable in Z-score units

Determining the correct size of a sample requires that we make certain decisions and guesses. For example, let's assume that we are drawing a sample of book costs from titles in our order file. We decide that we want to be accurate 95 times out of 100. Also, we are willing to permit a margin of error of $.20; that is, we will accept a mean cost that may be $.20 below or $.20 above the true mean cost. One further estimate must be made: the standard deviation. We may estimate it on the basis of experience or simple exploration of a few title costs. In our example, we estimate the standard deviation to be $1.50.

Calculating with our data in the formula above, we obtain:

$$n = \left\{ \frac{1.96\,(1.50)}{\$.20} \right\}^2 = 216 \text{ cases}$$

The 95 percent level of confidence (1.96 standard deviation units) is generally acceptable and may be used as a standard. More rigid confidence limits (such as the 99 percent level) would require a larger number of cases (1.57 standard deviation units would replace the 1.96 in the numerator). A more thorough explanation of confidence and confidence limits will be given in the following chapter.

In this chapter we have attempted to present, in an elementary fashion, the most frequently used sampling techniques. Once the variables of interest have been specified and the cases chosen, the data for those cases can be obtained. If the selected cases comprise the total population, descriptive statistics can be used to capsulize the information obtained in the form of summary statistics. If the cases are selected by a probability sampling technique, the data can be used to make estimates for the population as a whole on the variables of interest.

When we generalize from data gathered on a probability sample to the values on the variables for the whole population, we are using inductive statistics. Some of the theory and logic behind inductive statistics will be discussed in the following chapter.

REFERENCES AND SUGGESTED READINGS

Blalock, Hubert M., Jr. *Social Statistics*. 2d ed. New York: McGraw-Hill, 1972, p. 213–215.

Bookstein, Abraham. "How to Sample Badly," *Library Quarterly*, 44, no. 2: 124–132.

Mendenhall, William, Lyman Ott, and Richard L. Scheaffer. *Elementary Survey Sampling*. Belmont, Calif.: Duxbury, 1971.

Slonin, Mark. *Sampling*. New York: Simon and Schuster, 1960.

3
Inductive Statistics

The techniques in chapter 1 fall into a general class known as *descriptive* statistics. As we have noted, they summarize a body of data in several ways. They do not presume to go further than to describe the particular data under analysis.

Inductive statistical techniques give us a great deal more power—at the cost of certain risk-taking. Inductive statistics, also known as *inferential* statistics, is based on the classic purpose and logic of induction, that is, to generalize about a body of data on the basis of knowledge about a particular part or subset of that body of data. In short, induction means reasoning from the particular to the general.

For any number of reasons (usually because we lack adequate finances or because alternatives are not feasible), we are required to study a sample as a subset of the population we are interested in. The quantitative characteristics of samples are known as *statistics*. Samples are taken from populations in order to estimate the corresponding specific characteristics of that population. Because many different random samples can be taken from any population, statistics vary in their value from one sample to another.

A quantitative characteristic of a population is known as a *parameter*. Because all cases of the population are used in the calculation, a parameter has a fixed value that does not vary. In discussing samples and populations we use similar terms, for example, the mean, proportions, etc. However, when used to describe the sample, these terms have different symbols than when used for the population. The two that are most important for our purposes are the *mean*, which is denoted by $\overline{X}$ as a statistic of a sample and by μ (*mu*) as a parameter of a population, and the *standard deviation*, whose symbol as a sample statistic is s and is represented by σ (*sigma*) as a population parameter.

The Statistical Hypothesis

Generally, a hypothesis is a statement about an event or a set of events which are "unknown." In *hypothesis testing*, the investigator formulates a hypothesis, that is, makes certain assumptions about the values of or relationships between various parameters of the population. He or she must then decide whether the sample results contradict or support the assumptions. Hypotheses are essentially conjectured and are formulated in such a way as to make a formal statement about the values of a specific parameter(s) or about relationships among parameters in the population. In inductive statistics, hypotheses are of two sorts: the *research* hypothesis and the *null* or *statistical* hypothesis.

The research hypothesis consists of a guess or hunch about something, expressed as a declarative sentence. For example, we may hypothesize that *The level of education of library users is directly related to the extent of their use of the library*—and so formulate a research hypothesis, our hunch being that people with more education read more books.

Indeed, our hunch may be something we believe to be true, and consequently we often hear about "proving" a hypothesis. Without getting into the linguistic, metaphysical, and mathematical properties of proof, let us be certain of one cardinal principle. Statistical hypotheses are stated in such a way as to facilitate their rejection; hence the term test of hypotheses, rather than "proof." Remember, a test of a statistical hypothesis can never prove anything; it merely adds evidence which either supports or does not support our original theory. Honesty and rigor require that we do all we can to ensure objectivity throughout the process of collecting, measuring, and analyzing data—and, indeed, in conceptualizing our problem at the beginning.

The statistical or null hypothesis is a simple and straightforward means of "keeping us honest." Our null hypothesis for the example stated above would be that there is *no* relationship in the population between the level of education and the extent of library use. Of course, our hunch, guess, or belief (research hypothesis) may lead us to hope that this null hypothesis will be rejected. The logic of this approach is that we assume that no relationship exists. Then, if the evidence which accrues from our data is strong enough to indicate the existence of a relationship, we conclude that our hypothesis is false. That is, based upon our sample data, there is support for the theory that a relationship exists between the variables of interest in the population.

Moreover, following our test of the null hypothesis and its rejection or nonrejection, we must always ask: What else is true? A "good" hypothesis is one that virtually invites rival hypotheses. Our search for truth, for causation, for valid predictions of relationships between or among variables must follow a path of constant inquiry, of redefining and reconceptualizing, and of thinking up new alternative hypotheses about the problem or set of events under study. It is very important to consider what various possible conditions or intervening variables may affect our hypothesis.

We must account for three major sources or classes of error in statistical inference. First, no matter how carefully we attend to the mathematical reasoning and procedures, it is essential that the measurements we make are as reliable and valid as possible. The most adroit and precise statistical manipulation is of little value when our observations measures are inaccurate. Consequently, data collection and data processing must be carefully checked for error before we can safely proceed to data analysis.

Second, the nature of our sample and its processing must be appropriate for the inductive models we employ. In all instances in this book, independent random sampling is assumed for hypothesis tests and other inductive procedures. The characteristics of random sampling are discussed in the chapter on sampling.

Third, it is essential that we follow all other special requirements or assumptions for any given method. For instance, some hypothesis-testing models require that the distribution of the variable in the population follow a particular form. The t test requires that the variable have a normal distribution in the population. In addition, the t test should be conducted on data that are measured at the interval level in order to have interpretable results. The general chart of the statistical procedures, table 1, shows which tests are applicable for the various levels of measurement.

No mathematical rationale for matching measurement level and specific hypothesis-testing models will be offered here, but those who are interested will find ample discussion and explanation in the literature of mathematical and applied statistics (see suggested readings at the end of this chapter). Intuition alone will greatly help one see the necessity for selecting the right test, and the textual explication should go far to enhance understanding the essential correctness in selecting the proper test or measure. It is extremely important to employ the appropriate test or measure, lest all the labor of data collection be wasted in improper analysis.

Hypothesis testing involves the use of three interrelated distributions: the sample distribution (the distribution of the data in our sample), the sampling distribution (the theoretical distribution of a sample statistic), and the population distribution (the distribution of all the cases in the population on the variable of interest). The sampling distribution and the population distribution are theoretical, and although both can be constructed, we rarely have occasion to do so.

The most essential concept, however, in terms of understanding the logic behind the hypothesis-testing procedure, is the concept of the sampling distribution.

Imagine, for a moment, drawing a single sample of 5 library users and determining the *mean* number of books they have borrowed in the past year. Our "cases" here are individual library users and our variable of interest is number of books. The population of all library users is likely to be very large indeed, and would allow us to draw many samples from it. If we drew many samples, we would, each time, obtain a different mean number of books borrowed. There would be $\overline{X}_1, \overline{X}_2, \overline{X}_3,$ . . . —as many means as we have samples. These means may then be plotted as a distribution; as such, they would constitute a sampling distribution of means. If we were to ask *every* library user how many books he or she borrowed in the past year and tabulate the data, we could construct a distribution for the variable which would be considered a population distribution. Similarly, if we consider only the individuals in our original sample of 5 library users, and construct a frequency distribution for these data, we would have a sample distribution.

The hypothesis-testing procedure, of course, is not limited to means and sampling distributions of means. We can state and test statistical hypotheses about almost any statistic in which we are interested. If our statistical hypothesis concerns the population standard deviation, we will need to examine the sampling distribution of the sample standard deviation. If our interest lies in the population correlation, we will concern ourselves with the sampling distribution of the sample correlation coefficient.

Hypothesis Testing about the Population Mean

Let us recapitulate and look more closely at some of the major ideas in all hypothesis testing. First of all, after specifying the research hypotheses we plan our study so that we expect the results will not be affected by our sampling plan. Consequently, we use a random sampling plan in order to get the most representative sample.

A null hypothesis is stated about the population parameters and the sample data are collected and analyzed. For simplicity in the discussion that follows, let's assume that we are interested in evaluating the population *mean*, and that the null hypothesis states that the population mean, calculated on our variable, is 30. We now assume that our null hypothesis is true: that the population mean in fact equals 30. Even if the null hypothesis is true, sample statistics (sample means), calculated on a random sample taken from the population, will vary in value from sample to sample due to chance or random variation. This is because any selected sample will consist of fewer cases than the total population; therefore, the sample mean will be dependent on which cases are included in the sample we have selected. If we conceive of all possible values which the sample mean could equal, some of the values will be very likely, or will occur very frequently. In the case of the sample mean, a great proportion of the *sample* means will be very close to the true population means of 30. However, some of the values will occur very infrequently, or will be highly unlikely. We may get a few samples with means of 4 or 50.

A frequency distribution of all these possible sample means would portray which values are likely (relatively close to the population mean) and which values are highly unlikely (far from the hypothesized population mean). In defining the unlikely values, we usually (by convention) select those values which would occur about 1 or 5 percent of the time and are far from the hypothesized population mean. This total distribution of values, called a sampling distribution, is a theoretical distribution because we never take an infinite number of samples from a population.

We then take the value of our sample mean, calculated from the sample we randomly selected from the population, and compare it with the theoretical distribution of all sample means which could have been calculated from samples of the same size from the same population. If our sample mean takes a value we have labeled to be highly unlikely, in that it would rarely occur in a population with mean 30, we reject our null hypothesis. We are willing to believe that our null hypothesis is true until our sample evidence indicates otherwise.

We are, therefore, making a decision about our population parameter based on our sample statistic. In making this decision about whether to reject the null hypothesis, we take a certain risk of being wrong. The amount of risk we are willing to take is usually stated in terms of a percentage; the two conventional levels are 5 and 1 percent, but there is nothing sacred about either. This percentage, or risk we are willing to take, is associated with our definition of highly unlikely sample values which could occur under the null hypothesis. Remember, in reference to our previous example: Even though sample means of 4 or 50 are *un-likely* to occur when the population mean is 30, they *may* occur, though rarely.

Therefore, when our sample mean is far from our hypothesized population mean of 30, and values such as ours occur only about 5 percent of the time, our results are said to be significant and we conclude that our population mean is not equal to 30. But our decision may be wrong! If we were to repeat our hypothesis test infinitely many times, and the population mean *is* 30, we would reject a true hypothesis 5 percent of the time. Therefore, we would make the wrong decision 5 percent of the time. One of our objectives is to minimize the probability of our making this wrong decision and, thereby, rejecting a hypothesis which is in fact true.

Suppose the sample we originally took from our population consisted of 35 library users. Table 11 contains the *sample* distribution (or distribution of the values contained in our sample) for these users on the

Table 11. **Distribution of Number of Books Borrowed for Random Sample of Library Users**

No. of Books	Frequency
100	2
10	10
5	15
4	5
0	3

$$\overline{X} = [(2 \cdot 100) + (10 \cdot 10) + (15 \cdot 5) + (5 \cdot 4) + (3 \cdot 0)] / 35$$
$$= 395 / 35 = 11.3$$

number of books borrowed in the past year. The mean number of books borrowed for this group is 11.3. Figure 10 illustrates the frequency distribution of the total *population* of library users for the number of books borrowed; but notice that the true population mean has not been calculated. If our hypothesis is that the mean number of books borrowed last year for the total population was 30, the theoretical *sampling*

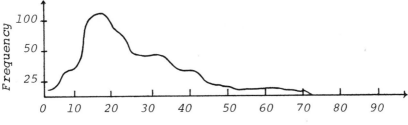

X = Number of books borrowed

Figure 10. Theoretical Frequency Distribution for Total Population of Library Users on the Variable Number of Books Borrowed

distribution (distribution of all possible sample means where $n = 35$) would look something like figure 11. Some samples of 35 individuals may have a mean as low as 6 books borrowed and some may have a mean as high as 40.

Looking at figure 11, we see that "unlikely occurrences" (or samples which result in means that are relatively far from the population mean) are found in the two tails of the sampling distribution. Values that occur in either the left or the right tail occur less often than values in the central portion of the distribution, which is why the height of the distribution at the tails is much less than that of the middle. We can specify the values in the tail regions which we consider to occur rarely by identifying the region cutoff points in the tails, determined by the shaded area in figure 11. This shaded area represents a certain proportion of the total area under the whole curve (or a certain proportion of the total number of samples in the population), and this proportion can be expressed in terms of a percentage (i.e., 5%; 2½% in

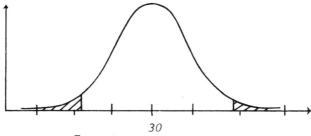

30

X̄ = Sample Mean Values
(Hypothesized Population Mean = 30)

Figure 11. Theoretical Sampling Distribution of the Mean Number of Books Borrowed by Library Users, Based on Random Samples of 35 Users

each tail). This area represents the region of risk, and the percentage represents the times (in the long run) that we would reject a true hypothesis.

Remember, we are assuming that we have identified the true population mean, and the sampling distribution is determined by our null hypothesis (which will be explained when we cover the central-limit theorem). Therefore, even when the null hypothesis is true, we sometimes get sample values which are quite different or far from the true population mean, and because they are so "far," we will incorrectly reject the null hypothesis.

Up to this point we have been vague about identifying values which we consider to be rare or unlikely. We can do this by choosing our sample size and risk or significance level first. Then, through the information given to us in the central-limit theorem, we can use Z scores to determine the cutoff values in the shaded areas. Let us examine the central-limit theorem and then go through the whole procedure step by step by means of an example.

The Central-Limit Theorem

You may have noticed that the sampling distribution of the mean in figure 11 centers around the hypothesized population mean of 30, and seems to be a normal distribution. The distribution in figure 11 is pictured that way because of the information given to us by the central-limit theorem. This remarkable theorem gives us the "legitimate" basis for using the normal distribution as an inductive model when (1) we are testing hypotheses concerning a *population mean* and (2) our sample size is relatively large. We are perfectly capable of testing hypotheses about other population parameters, or relations between population parameters, but the mathematical models or theoretical sampling distributions which we would use in testing the hypothesis might not follow a normal distribution. At this point we are concerned only with a hypothesis about the population mean.

If we draw as many independent random samples as we can of a specific size from a population with mean μ and variance σ^2, we can form a sampling distribution of sample means that is calculated from these independent random samples. The central-limit theorem states that if the sample size is relatively large (i.e., greater than 30), the sampling distribution of the mean $(\overline{X})$ becomes approximately or almost normally distributed, with mean μ and variance σ^2 / n. The larger the sam-

ple size, the closer the sampling distribution to being normally distributed. Note that the sampling distribution of the mean for relatively large
samples will be approximately normally distributed, *no matter what the
shape of the original population distribution!* With this information, we
do not need to know the shape of the distribution of the population
on the variable which is of interest to us. Regardless of the shape, we
can assume that the theoretical sampling distribution will be almost
normal in form. Therefore, we can use information about the family
of normal distributions to determine what percentage of the total area
under the curve is in the tails of the distribution, which we have men-
tally shaded or consider to be unlikely sample mean values.

We can now be more specific about illustrating the sampling distribution in figure 11. The central-limit theorem tells us that for a population
with mean μ and variance σ^2, *if* n is relatively large, or greater than 30,
the sampling distribution of samples of size n will be approximately
normally distributed with mean μ and variance σ^2 / n. Our sample size
was 35; therefore our sampling distribution will be nearly normal. Also,
our hypothesized population mean was 30; therefore the sampling distribution will be centered around (have a mean of) 30. This fact—that
the mean or central point of the sampling distribution of the mean
equals μ—is equivalent to stating that the sample $\overline{X}$ is an *unbiased*
estimator of the population mean μ. The most likely value for one
sample mean, which we would obtain by repeated sampling, would, in
the long run, be equal to μ.

Now we would like to determine the variability of the sample mean, or
the variance of the sampling distribution of the mean $(\overline{X})$. The central-
limit theorem indicates that for a population with a population variance
σ^2, the variance of the sampling distribution of $\overline{X}$ for samples of size n,
will equal σ^2 / n. We do not generally know the true population variance,
σ^2; therefore we are forced to use our best estimate of it. Our best
estimate would be our sample variance, s^2. We defined s^2 in chapter 1 as

$$s^2 = \frac{\sum_{i=1}^{n}(X - \overline{X})^2}{n}$$

At that point we were discussing descriptive statistics. We are now
concerned with being able to make inferences; so we will have to change
our formula for s^2 slightly. We stated that the sample mean is an unbiased estimator of the population mean, or that $\overline{X}$ is an unbiased

estimator of μ. The sample variance is a *biased* estimator of the population variance, or s^2 is a *biased* estimator of σ^2. An estimator is biased if, on the average, it is not equal to the population parameter it estimates. The central point (mean) of the sampling distribution of the sample variance s^2 is *not* equal to σ^2 but, rather, is *smaller* than σ^2. We would like to correct for this bias; so we use the following formula for the sample variance, s^2, whenever we use inductive statistics:

$$s^2 = \frac{\sum_{i=1}^{n} f(X - \overline{X})^2}{n - 1}$$

Since we are dividing by a smaller number $(n - 1)$, we will get a value which will be larger, and in fact an unbiased estimator of the true population variance.

Returning to the variance of the sampling distribution of the mean, we can calculate the sample variance of the data in table 11:

X	f	$\overline{X}$	$(X - \overline{X})$	$(X - \overline{X})^2$	$f(X - \overline{X})^2$
100	2	11.3	88.7	7,867.69	15,735.38
10	10	11.3	−1.3	1.69	16.90
5	15	11.3	−6.3	39.69	595.35
4	5	11.3	−7.3	53.29	266.45
0	3	11.3	−11.3	127.69	383.07
					16,997.15

$$s^2 = \frac{\Sigma f(X - \overline{X})^2}{n - 1} = \frac{16,997.15}{34} = 499.92$$

An f appears in the formula because we have more than one case for each X value which occurs. If we did not multiply by the frequency of occurrence, we would have indicated only 5 instead of 35 individuals.

With this information, we can determine the variance of the sampling distribution:

Estimate of $\sigma^2 / n = s^2 / n = 499.92 / 35 = 14.28$

What we are really interested in is the estimate of the standard deviation of the sampling distribution of the mean, $\overline{X}$, which is the square root of its variance. $\sqrt{\sigma^2 / n}$ is also referred to as the *standard error of the mean* and has a special symbol, $\sigma_{\bar{x}}$. Therefore:

Estimate of $\sigma_{\bar{x}} = \sqrt{14.28} = 3.78$

Putting together all this information, with the knowledge that the distribution is nearly normal, we can construct figure 12, which is an illustration of the sampling distribution of $\bar{X}$ for a population with μ

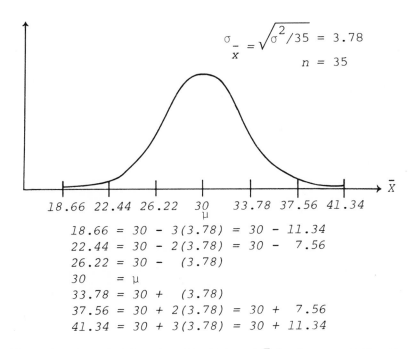

$$\sigma_{\bar{x}} = \sqrt{\sigma^2/35} = 3.78$$

$$n = 35$$

18.66 22.44 26.22 30 33.78 37.56 41.34
                        μ

$$18.66 = 30 - 3(3.78) = 30 - 11.34$$
$$22.44 = 30 - 2(3.78) = 30 - 7.56$$
$$26.22 = 30 - (3.78)$$
$$30 = \mu$$
$$33.78 = 30 + (3.78)$$
$$37.56 = 30 + 2(3.78) = 30 + 7.56$$
$$41.34 = 30 + 3(3.78) = 30 + 11.34$$

Figure 12. Theoretical Sampling Distribution of $\bar{X}$ for Samples of Size 35 Taken from Population with $\mu = 30$ and Estimated Population Variance $\sigma^2 = 499.92$

equal to 30, based on samples of size 35. The values of the sample means, indicated along the horizontal axis (18.66, 22.44, etc.), are calculated as indicated at the base of figure 12.   18.66 is 3 standard deviations away from the mean, 30, as is 41.34.   22.44 and 37.56 are each 2 standard deviations away from the mean, 30.   26.22 and 33.78 are each 1 standard deviation away from the mean, 30. We can conceive, then, of points 18.66 and 41.34 as 3 standard deviation units away from the mean. This means that 18.66 has a Z score of -3.0, since it is 3 standard deviation units *below* the mean. Similarly, 41.34 has a corresponding Z score value of 3.0, since it is located 3 standard deviation units *above* the mean.

All Z score values can be determined as follows.

$\overline{X}$	Z	Explanation
18.66	−3.00	3 standard deviations below the mean
22.44	−2.00	2 standard deviations below the mean
26.22	−1.00	1 standard deviation below the mean
30.00	0.00	the population mean
33.78	1.00	1 standard deviation below the mean
37.56	2.00	2 standard deviations above the mean
41.34	3.00	3 standard deviations above the mean

Figure 13 illustrates this Z score transformation.

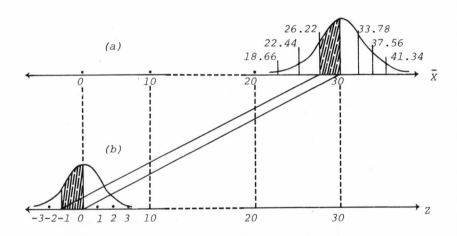

Figure 13. Z Score Transformation from a Sampling Distribution with $\overline{\mu} = 30$ and $\sigma_{\overline{x}} = 3.78$ into a Unit Normal Distribution with $\mu = 0$ and $\sigma = 1$. Shaded Areas in Both Distributions Represent 34% of Total Area under the Curve

We are transforming our theoretical sampling distribution, with μ equal to 30 and standard error of the mean ($\sigma_{\overline{x}}$) equal to 3.78, into a "unit normal distribution," which is a normal distribution, with mean 0 and standard deviation equal to 1.0. We are changing the distance between the specified points from 3.78 units to 1 unit. We are also shifting the mean from 30 to 0. The proportion of the total area between any two of the specified points will remain the same because each of these distributions is normal! This is one of the special properties of the normal distribution which we discussed in chapter 1. *Any* point in the sampling distribution in figure 12 can be transformed into a Z score through the following formula:

$$Z = \frac{\overline{X} - \mu}{\sigma_{\overline{x}}}$$

This is conceptually similar to the formula at the end of chapter 1, in that we subtract a mean (the mean of the means) and divide by a standard deviation. By using this formula we are shifting each score 30 units to the left and, also, reducing the standard deviation from 3.78 to 1. The percentage of total area under one normal curve that is bounded by two values will be the same as the percentage of total area under another normal curve bounded by the corresponding two values (i.e., the values obtained by transforming the initial two values). For instance, 34 percent of the total area under curve *a* lies between the values 26.22 and 30.00. By using our Z score transformation, we find that the value 26.22 corresponds to −1.00 on our new curve (b) and 30.00 corresponds to 0.00. The area bounded by −1.00 and 0.00 on curve *b* is 34 percent of the total area under the curve. Normally, we would need to use calculus to calculate the area under *a* curve. However, in the special case of curve *b*, the unit or standard normal curve, the calculations have been performed for us and the results tabulated in tables like number 2 in the Appendix.

Testing a Hypothesis about μ When *n* Is Large

Let us now return to our original sample data, based on 35 library users, and test our hypothesis that the population mean is equal to 30.

Our sample data resulted in an average 11.3 books borrowed last year. Is this a likely or unlikely sample mean value, if our population mean is assumed to be equal to 30?

First, we must specify the amount of risk we are willing to take in rejecting a hypothesis which is in fact true. We will set this percentage of risk, called the significance level, at 5 percent. If our sample mean lies in the area of rejection in the tails of the sampling distribution which encompasses 5 percent of the total sample means, we will be willing to say our findings are highly unlikely to occur under a population mean of 30, and therefore we will reject our null hypothesis. Rather than the real sampling distribution in figure 12, we intend to use the unit normal distribution. We must find the Z scores located at the beginning of the shaded areas in figure 14 in order to make our decision.

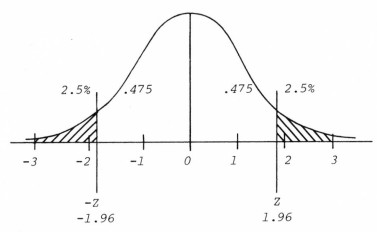

Figure 14. Unit Normal Curve with Shaded Area Which Indicates Unlikely Sample Values Which Would Cause Us to Reject Our Null Hypothesis

We will transform our sample mean of 11.3 into a Z score and see if it falls into the shaded region of figure 14. If it does, we will reject our hypothesis that the population mean equals 30. We split our 5 percent (or .05) into two equal parts and place half in each tail: .025 or 2.5 percent in the left tail and .025 or 2.5 percent in the right tail. This indicates that if we get a very small *or* a very large sample mean, we will reject our hypothesis. The area under the curve in figure 14, from 0 to Z, equals .475 or 47.5 percent of the total area (since the total area to the right of 0 equals .50 or 50 percent of the total area).

At the center of appendix table 2, with the value .4750, we see that the row value of the Z is 1.9 and the column value of the Z is .06. Adding these together, we have a Z score of 1.96, which means that the area between a Z value of 1.96 and the mean 0 equals .4750. Since the unit normal distribution is symmetric, we know that $-Z$ equals -1.96. Therefore, our *region of rejection* (shaded area) in figure 14, which will cause us to reject our hypothesis, consists of all sample mean values which result in Z scores *less than* -1.96 or *greater than* 1.96.

Our sample mean results in the following Z score:

$$Z = \frac{\overline{X} - \mu}{\sigma_{\bar{x}}} = \frac{11.3 - 30}{3.78} = -4.95$$

-4.95 lies to the left of -1.96 and is *inside* the shaded region of rejection. Therefore, since getting a sample with a mean of 11.3 seems unlikely when the population mean is 30, we conclude that our hypothesis is false. We decided that our assumption that the population mean equals 30 was not correct. (Remember, we are taking a risk of being wrong when we make this decision.) We would conclude, based upon our data, that the population mean is less than 30. At this point, our best estimate of the population mean is 11.3.

In review, to test a hypothesis about the *population mean* we go through the following steps.

1. Make an assumption about the value of the population mean and state it as your null hypothesis.
2. If the sample size is greater than 30, we can use the unit normal distribution to help us make our decision, since the central-limit theorem indicates that the sampling distribution of the mean will be approximately normally distributed in this case.
3. Select a risk or significance level.
4. Find the Z scores by using appendix table 2, which corresponds to the region of rejection in the tails of the distribution. (We recommend drawing a picture of the distribution and shaded region of rejection.)
5. Take a random sample and calculate the sample mean and sample (unbiased) standard deviation. Determine the standard error of the sampling distribution of the mean for samples of this size.

$$\text{Estimate of } \sigma_{\bar{x}} = \sqrt{\frac{s^2}{n}}$$

6. Using your sample information, calculate the sample Z value (test statistic) by the following formula:

$$Z = \frac{\overline{X} - \mu}{\sigma_{\bar{x}}}$$

7. Make the decision. If the sample Z value falls in the region of rejection, reject the null hypothesis. If it does *not* fall in the region of rejection, *tentatively* assume that the null hypothesis is true. Relate the findings to your research hypothesis.

Testing a Hypothesis about μ When n Is Small

The central-limit theorem applies to the sampling distribution of the mean only when the sample is relatively large. When the sample is relatively small, the sampling distribution of $\overline{X}$ does not follow a normal distribution but, rather, a t distribution, often referred to as "the student's t distribution." The t distribution, another special mathematical distribution, serves in problems of testing hypotheses about means when the sample is small (usually less than 30). We said in the previous section, however, that the sampling distribution of the mean will be *approximately* normally distributed when n is greater than 30. It becomes *truly* normally distributed only when n is extremely large ($n >$ 100). The true distribution before this is the t distribution. Using the t distribution in testing hypotheses about means where $n < 100$ will be more exact, since the sampling distribution of the mean exactly follows a t distribution.

The t distribution is slightly differently shaped than a normal distribution; it is necessary to go a larger number of standard deviation units from the mean to include, for instance, 95 percent of the area or cases under the curve. As the number of cases (n) grows larger, the t distribution looks more and more like the normal distribution, as examination of the t table will show (see appendix table 3).

Just as there is a family of normal distributions, there is a family of t distributions, with similar characteristics. All have a mean of 0 and each t distribution has, associated with it, a parameter called "degrees of freedom," indicated by the symbol df. Ordinarily, in most applications of the t distribution to problems that involve a single sample, $df = n - 1$, or 1 less than the number of independent observations in the sample. The concept "degrees of freedom" deals with the number of values that are free to vary.

If we specify a mean of $\overline{X} = 10$ and we know that there are 4 cases in our sample, we are free to choose the first three values ($X_1 = 5$; $X_2 = 20$; $X_3 = 10$). This sums to $5 + 20 + 10 = 35$, which means that X_4 must equal 5 in order to result in a mean of 10 ($\overline{X} = (5 + 20 +$

$10 + 5) / 4 = 10$). There are, therefore, three degrees of freedom, because three values are allowed to vary.

To test a hypothesis about a population mean by using the t distribution, the procedure is similar to that specified in the previous section, with one important difference. For the sampling distribution of the mean to follow a t distribution for $n < 30$, the original population from which the sample was taken *must have a normal distribution* for the variable in which we are interested. We do not need this assumption with relatively large samples, because the central-limit theorem takes effect. With small samples, we need this extra assumption.

The region of rejection for testing hypotheses about means is found, as before, by entering the table with the significance level or amount of risk we are willing to take. We will, however, be entering appendix table 3, which is the t distribution table, and we will need to know the df value $(n - 1)$ to get the specific points on the edge of our shaded region of rejection. An example of a test of a hypothesis will illustrate the similarity to the previous procedure.

A recent report informs us that the mean price of reference books this year is reported to be \$20. We find that the mean price of a random sample of reference books purchased by our library is \$24, with a standard deviation of \$3. Given that our 16 titles are only a sample of all reference titles, we formulate the null hypothesis that there is no significant difference between the average price of books in our sample and those in the total reported group. Specifically, our null hypothesis is that the population mean equals \$20. To test our hypothesis, we must first make the assumption that the population distribution of reference-book prices is normal, which seems a fairly reasonable assumption. Because of our sample size, we must test this hypothesis by a t test.

Let us assume that we select a significance level of 5 percent. We must use appendix table 3 to find the t values on the borders of our region of rejection, shown in figure 15. As before, we divide our .05 into two equal parts and enter table 3 with an area of .025. Looking across the row where $df = 15$ and down the column marked .05, we arrive at a t value of 2.131. The t table includes only positive t values, but because of the symmetry of the distribution we know that the t value at the edge of the shaded area in the left tail must equal -2.131. Using our sample information, we calculate the standard error of the mean, which will equal, as before:

$$\text{Estimate of } \sigma_{\bar{x}} = \sqrt{\frac{s^2}{n}} = \sqrt{\frac{9.00}{16}} = .75$$

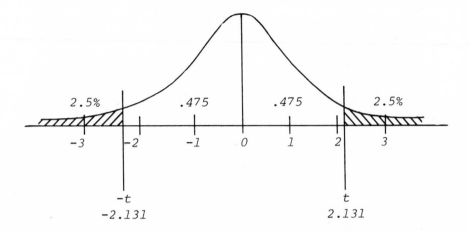

Figure 15. Student's *t* Distribution for (*n* − 1) = 15, with Shaded Area Which Indicates Region of Rejection for Our Null Hypothesis

We now calculate our sample test statistic, which in this case will be a *t* value. The formula is similar to the *Z*-score formula:

$$t = \frac{\overline{X} - \mu}{\sigma_{\bar{x}}}$$

The *t* value for our sample mean is

$$t = \frac{24 - 20}{.75} = 5.33$$

This value falls inside the region of rejection, which means that it is far enough from our hypothesized population mean of 20 for us to conclude that our hypothesis is incorrect. We have statistically significant results, which makes us question whether our sample came from a population with a mean of $20. Again, we have two basic ways of testing null hypotheses about the population mean: the normal distribution or the *t* distribution. When the sample is fairly small, we should use a *t* test, if the assumption of normality in the population distribution seems reasonable. If the sample is large, we can still use a *t* distribution, if we wish to be very exact, but the *t* distribution becomes so nearly normal in form that it does not make much difference. Also, when the sample is relatively large, whether we use t or Z we need not assume normality in the population distribution, since, with large samples, the sampling distribution of the mean will approach normality anyway.

Confidence Limits

In drawing a sample and calculating its mean, our best estimate of the population is the sample statistic. We know that repeated samplings would produce various means, but what we want to know is how reliable our sample mean is. We can do this by calculating *confidence limits* or a *confidence interval*.

Confidence limits consist of a point above and a point below the sample mean. In setting these limits, we are saying that we expect repeated sample means to fall within them. The usual confidence coefficient is .95, by which we assume that if we were to repeat our sampling an extremely large number of times, 95 times out of 100 we would expect the sample means to fall within the limits or interval.

Assume that we have a sample $n = 100$, with a mean of 50 and a standard deviation of 15. At the .95 level, we simply multiply the standard deviation by 1.96. The area under the unit normal curve bounded by the values -1.96 and 1.96, you will recall, includes 95 percent of the total area under the curve.

The general formula for establishing confidence limits is

$$\overline{X} \pm 1.96\,\sigma_{\bar{x}}, \text{ where we estimate } \sigma_{\bar{x}} = \sqrt{\frac{s^2}{n}}$$

In our problem, this becomes

$$50 \pm 1.96\,\sqrt{\frac{(15)^2}{100}} = 50 \pm 2.94 = 47.06, 52.94$$

We would expect, then, that if we were to take a large number of samples, 95 times in 100 their means would fall between 47.06 and 52.94. If we want to use 99 percent confidence limits, we would substitute 2.58 for 1.96. The distance between the upper and the lower limit gives us an indication of how reliable our sample mean is. A distance of .05 would mean that the sampling distribution of the sample mean is very narrow and, also, that the sample means are closely grouped together. A distance of 50, however, would indicate that the sample means have great variability and are widely scattered around the population mean. In constructing confidence limits for samples which have few cases, we use the t value instead of the Z value, just as we did in hypothesis testing.

Difference of Means Test

A more common use of the t test occurs when we have two independent random samples for normally distributed population and we wonder if the differences between the sample means are due to chance or are statistically significant. In other words, are the two samples from two populations with equal means or are they from two populations with two distinct population means? In stating our null hypothesis, we assume that the two population means are equal and that the two population variances are also equal. Assume that we take two independent random samples of books from the physical and social sciences, and we have computed their average prices with the following results (notice that we have small samples):

Physical Sciences	Social Sciences
$\overline{X}_1 = \$17.90$	$\overline{X}_2 = \$16.40$
$s_1 = 2.30$	$s_2 = 2.10$
$n_1 = 20$	$n_2 = 30$

Our question is: Are the prices of Physical and Social Science books significantly different?

In testing this, we make the assumption that they are not, that the *population* mean prices are not different (the null hypothesis). We could compute the difference between the two means for this example, and it would not be equal to 0. If the two population means are equal, the difference between them would be equal to 0. If the two population means are equal, imagine taking two independent random samples, one from each population. Sometimes we would get a positive difference between the sample means, sometimes we would get a negative difference between the sample means. Most likely, however, we would get no difference between the two sample means.

We could imagine taking as many pairs of samples as possible from the two populations and determining the differences between the sample means. We could form a distribution of these differences, which would be a sampling distribution of the difference between two sample means. If the null hypothesis is true, this sampling distribution will be distributed as t with $n_1 + n_2 - 2$ degrees of freedom, with mean 0 and standard error (of the difference between means) equal to $\sigma_{\bar{x}_1 - \bar{x}_2}$:

$$\text{Estimate of } \sigma_{\bar{x}_1 - \bar{x}_2} \sqrt{\frac{s_1{}^2}{n_1} + \frac{s_2{}^2}{n_2}}$$

Selecting the .05 level of significance, we can determine the region of rejection, as before, by looking up the scores in appendix table 3 which correspond to this area. The closest t scores for the .05 level with 48 degrees of freedom are -2.02 and 2.02. We will transform our sample difference into a t score, and if it is less than -2.02 or greater than 2.02 we will reject our null hypothesis. Figure 16 contains the sampling distribution of the difference between means, based

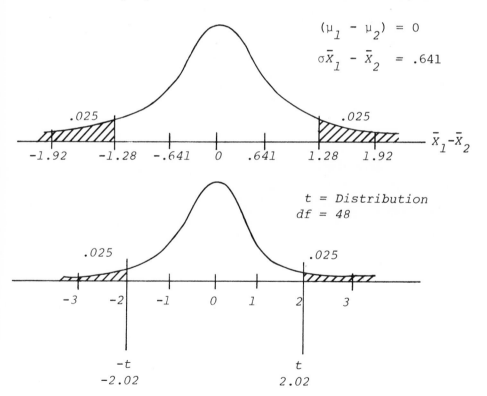

Figure 16. Sampling Distribution of Difference between Sample Means Based on Samples of Size $n_1 = 20$ and $n_2 = 30$ and Corresponding t Distribution

upon our samples, and the corresponding t distribution with the appropriate region of rejection shaded. To transform our sample difference into a t score, we use the following formula:

$$t = \frac{(\bar{X}_1 - \bar{X}_2) - (\mu_1 - \mu_2)}{\sigma_{\bar{x}_1 - \bar{x}_2}}$$

Once again we have a score, from which we subtract its population mean (the population means are equal, therefore their difference will be 0), and divide by the standard error or standard deviation of the sampling distribution. Our sample t value equals

$$t = \frac{(17.90 - 16.40) - 0}{\sqrt{\dfrac{(2.30)^2}{20} + \dfrac{(2.10)^2}{30}}}$$

$$= \frac{1.5}{\sqrt{.265 + .147}}$$

$$= \frac{1.5}{.641}$$

$$= 2.34$$

Our computed results allow us to reject the null hypothesis and conclude that the prices are significantly different at the .05 level.

We would expect that results as large as ours would not occur more than 5 times in 100. Note, however, that our computed t score would not be significant at the .01 level had we elected it as our significance level. Results of the size we have would be expected to occur by chance alone between 1 and 5 times in 100.

As a second example of how the differences of means test work, consider the following. Our attention has been drawn to the possibility of paying significantly different salaries to beginning librarians. Presumably, males have been paid more than females. Our sample study of salaries for the past 5 years shows the following results.

Males	Females
$\bar{X}_1 = \$10,600$	$\bar{X}_2 = \$9,800$
$s_1 = 1,200$	$s_2 = 1,300$
$n_1 = 25$	$n_2 = 36$

That is, our study of 25 males and 36 females shows that males were paid, on the average, $10,600 and females $9,800. Their respective standard deviations were $1,200 and $1,300. We make the following assumptions:

1. We have taken independent random samples.
2. The populations from which we sampled were normally distributed with equal variance.

It also makes sense that the variable under consideration, in calculating means, was measured at the interval level. Our null hypothesis is that there is no significant difference between males' and females' salaries. We are thus hypothesizing that the population mean of male and female salaries is the same. We select a .05 level of significance. This means that we are willing to risk being wrong in rejecting the null hypothesis 5 times in 100 if we were to draw a very large number of samples.

$$t = \frac{(\overline{X}_1 - \overline{X}_2) - 0}{\sqrt{\dfrac{s_1^2}{n_1} + \dfrac{s_2^2}{n_2}}}$$

$$= \frac{800}{\sqrt{\dfrac{(1,200)^2}{25} + \dfrac{(1,300)^2}{36}}} = \frac{800}{\sqrt{57,600 + 46,944}} = \frac{800}{323.33}$$

$$= 2.47$$

$$df \text{ (degrees of freedom)} = n_1 + n_2 - 2 = 59$$

Consulting the t table, we find that our results are significant at the .05 level—that we can reject the null hypothesis that there is no significant difference between the population mean salaries for male and female librarians. That is, if the population mean salaries for males and females are equal, we would be likely to obtain such a large difference, as the one we found, in only 5 percent of the samples we could take. Therefore, we conclude that the population mean salary for males is higher than the population mean salary for females.

Again, we may have made an error in our decision. Only further research will indicate whether our findings are a true indication of the real state of affairs.

In this chapter we have presented the basis of hypothesis testing by concerning ourselves with tests about population means. The Z and t distributions, as used here, are applicable for tests involving one or two sample means. Other statistical hypotheses may be tested about population parameters other than the mean, but the mathematical distributions that are used for sampling distributions may not be a normal or a t distribution. Nevertheless, these tests are so common that we thought it best to present them in this chapter. References at the end of this chapter will cover other kinds of tests about sample means or between sample means which do not meet the assumptions or conditions which we presented here.

We now turn to the relationships between variables in a population.

Exercise

1. Some people in the library argue that things have changed in respect to salaries in the past year. Again, we take samples of men and women beginners' salaries for the past year and obtain the results below. Using the t test, what can we say about the presumed changes?

Males	Females
$\bar{X}$ salary of \$10,750	$\bar{X}$ salary of \$10,491
$s_1 = \$150$	$s_2 = \$200$
$n_1 = 5$	$n_2 = 8$

REFERENCES AND SUGGESTED READINGS

Blalock, Hubert M., Jr. *Social Statistics*. 2d ed. New York: McGraw-Hill, 1972, p. 188–193, 219–228.

Glass, Gene, and Julian C. Stanley. *Statistical Methods in Education and Psychology*. Englewood Cliffs, N.J.: Prentice-Hall, 1970.

Hays, William. *Statistics for the Social Sciences*. New York: Holt, Rinehart and Winston, 1973.

4
Correlation and Regression

In previous sections we have seen that statistical tests of significance tell us whether there is a relationship between variables. For example, in chapter 3 we found that, based on our two independent random samples, there seemed to be a relationship between sex and yearly salary for librarians. The males in our sample received a significantly higher mean salary than the females.

Often we are more interested in knowing how *strong* the relationship is between variables and how we may "predict" such a relationship. A standard and important way to obtain such information is by using procedures for correlation and regression analysis.

Correlation

The major procedure to be considered in this section is known as *Pearson's product–moment correlation*, or *zero-order correlation*, symbolized by the letter r. It is expressed as a decimal, ranging in value from minus 1 to plus 1 (-1 to $+1$). Any given value for r indicates the strength of linear association between variables. A large positive or negative value of r shows strong association between variables. The association may be positive: as one variable increases, the other increases linearly; *or* it may be negative: as one variable increases, the other decreases linearly. Such an association should never be considered *causal* without evidence other than that ordinarily provided by the correlation analysis itself.

For example, a given independent variable, X, such as library budget, is commonly associated quite powerfully with the variable Y, size of library collection. A "causal" connection might be asserted by reasoning that the more money a library staff has to spend, the more materials it will necessarily obtain $(X \rightarrow Y)$. Such reasoning may be plausible, but no causal connection can be legitimately claimed. Actually, it might well be argued that the two variables are interdependent $(X \longleftrightarrow Y)$, or that one or more other variables (Z), such as user needs and willingness to pay for these needs, determine or cause both the size of the budget and the size of the collection.

$$\text{or } Z \rightarrow X \rightarrow Y$$

Correlation can be shown to be a function of regression. Regression analysis is basically a particular way of stating a "law" in the scientific sense, for it provides the basis for predicting the value (s) of one variable from the value (s) of another. This familiar equation is the mathematical formulation for regression analysis involving two variables:

$$Y = a + bX$$

With two-variable regression we try to predict the values of one variable (Y), given those values of another variable (X). We try to come as close as possible to discovering a "law" between the two variables, but, inevitably, the relationship will not be perfect and error in our predicted values of Y will occur. Therefore, the equation should be restated as

$$Y = a + bX \pm \text{error}$$

This formula says, in effect, that once a solution for a and for b is derived, a value for the variable Y can be predicted for any given value for the variable X. Regression analysis is usually multivariate in nature. We usually want to predict values of Y, given values of many different X variables. However, to understand the general nature of regression it is best to start with simple regression analysis.

Scattergrams

When we are interested in seeing if one variable is associated with another, we draw a *scattergram*, placing each case in a position according to its X and Y values. A scattergram is a two-dimensional graph made up of points whose coordinates are determined by the two variables under study. Figure 17 is a scattergram based on table 12 that shows the pattern of association between per capita expenditure and per capita circulation in a sample of 10 branch libraries. The base line or abscissa measures the per capita expenditure; the ordinate or vertical axis marks the per capita circulation. Branch A, for instance, has a per capita expenditure of $4.60 and a per capita circulation of 4.5 books.

The scattergram is extremely important in showing us the nature of the relationship; it will show the *direction* of the relationship, whether it is negative or positive, and whether the relationship is *linear*. In figure

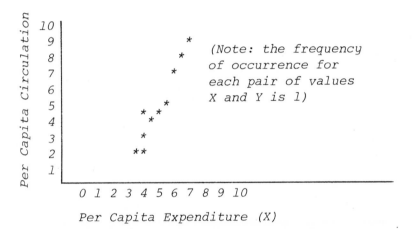

(Note: the frequency of occurrence for each pair of values X and Y is 1)

Figure 17. Scattergram Containing Data Points for 10 Libraries on 2 Variables

Table 12. **Data for 10 Branch Libraries Measured on 2 Variables: per Capita Expenditure (X) and per Capita Circulation (Y)**

Case	Variable X	Variable Y
A	$4.60	4.5
B	4.10	4.6
C	6.70	8.2
D	3.90	2.5
E	3.00	2.1
F	5.10	4.9
G	4.00	3.9
H	7.20	8.9
I	6.20	7.7
J	5.50	5.4

17 the relationship is positive, for as per capita expenditure increases, so does per capita circulation. Further, we may visualize a straight line almost underlying the points; thus the relationship is linear. A linear relationship between two variables Y and X is one in which Y is mathematically dependent on X; that is, Y depends on X, not X^2, X^3, etc. If we graph the r values of X and Y that fit the equation $Y = a + bX$, the results will always be a straight line, although our graph will show different lines depending on the values of a and b we use.

The usefulness of the product–moment correlation requires that we can safely assume a linear relationship. Figure 18 shows an interesting

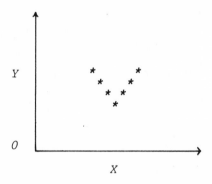

Figure 18. Scattergram Showing Relationship between 2 Variables for Which Pearson r Is Not an Appropriate Statistic

relationship, but it is curvilinear, not linear, and the procedure for r is not appropriate.

A scattergram may also show that there is no important association, linear or otherwise, as in figure 19. For any given value of X there are many values of Y, and vice versa.

The scattergram in figure 20 demonstrates a problem that arises when a study has not provided an adequate sample. The study data are indicated in the area outlined by the rectangle. If the study had included more data or a more representative sample, it would include the cases outside the rectangle. In such a case, we would incorrectly assume that there is little or no association between the variables on the basis of the study data—when further study would show a fairly important association.

When there are relatively few cases, it will be quite simple to see if there is some kind of relationship—whether one variable "goes with

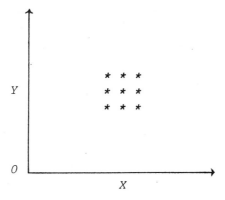

Figure 19. Scattergram Illustrating No Association between Variables *X* and *Y*

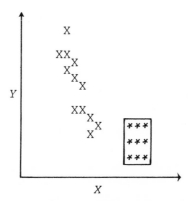

Figure 20. Scattergram Illustrating Nonrepresentative Sample Which Results in Poor Estimate of the Correlation between *X* and *Y*

another." Scanning the *X* and *Y* columns of figures in table 12, we see that as per capita expenditures increase, per capita circulation increases too. When there are many cases, it is difficult to guess the path of the data with any accuracy; hence the importance of the scattergram, which neatly summarizes the data in one simple picture.

Scattergrams can be produced fairly simply on graph paper by hand. However, more often than not you will have data that are machine-readable, and producing a graph by computer is a simple matter with a "canned" computer program such as Statistical Package for the Social Sciences (SPSS). Requesting a printout of a scattergram with the numerical analysis (*r*) will enable you to make a more reliable and valid interpretation of the results.

Calculating r

The steps for calculating a correlation coefficient are direct and uncomplicated, as the following example shows. Performing the calculations without a computer can be cumbersome if the number of cases is large. Given the wide availability of machine data processing, it is usually very advantageous to use the computer. It will produce results more readily, and can produce a scattergram as well. The general formula for r is

$$r = \frac{n\Sigma XY - (\Sigma X)(\Sigma Y)}{\sqrt{[n\Sigma X^2 - (\Sigma X)^2][n\Sigma Y^2 - (\Sigma Y)^2]}}$$

Example: What is the correlation between per capita expenditures and per capita circulation among the 10 branch libraries in table 13?

$$r = \frac{n\Sigma XY - (\Sigma X)(\Sigma Y)}{\sqrt{[n\Sigma X^2 - (\Sigma X)^2][n\Sigma Y^2 - (\Sigma Y)^2]}}$$

$$= \frac{10\,(292.66) - (50.30 \times 52.70)}{\sqrt{[10\,(269.61) - (50.30)^2][10\,(326.19) - (52.7)^2]}}$$

$$= \frac{2{,}926.60 - 2{,}650.81}{\sqrt{(2{,}696.10 - 2{,}530.09)(3{,}261.90 - 2{,}777.29)}}$$

$$= \frac{275.79}{\sqrt{(166.01)(484.61)}} = \frac{275.79}{\sqrt{80{,}450.106}} = \frac{275.79}{283.64} = .972 = .97$$

The value of r ranges from $+1$ to -1. As our results produced an r of .97, a strong positive association exists between the variables. The amount of change in Y is very strongly associated with a change in X. More specifically, an increase in X is accompanied by a commensurate *increase in Y*; therefore, the association is *positive*.

Recall that it is very possible to have a powerful association that is *negative*. That is, *increase* in X would be accompanied by *decrease* in Y. Among adult users of public libraries we may find such an association: as adults grow older, they use the library less.

Table 13. **Data Gathered from 10 Branch Libraries on 2 Variables: per Capita Expenditure and per Capita Circulation**

Branch Library	Per Capita Expenditure	X²	Per Capita Circulation	Y²	XY
A	$ 4.60	21.16	4.5	20.25	20.70
B	4.10	16.81	4.6	21.16	18.86
C	6.70	44.89	8.2	67.24	54.94
D	3.90	15.21	2.5	6.25	9.75
E	3.00	9.00	2.1	4.41	6.30
F	5.10	26.01	4.9	24.01	24.99
G	4.00	16.00	3.9	15.21	15.60
H	7.20	51.84	8.9	79.21	64.08
I	6.20	38.44	7.7	59.29	47.74
J	5.50	30.25	5.4	29.16	29.70
Total	$50.30	269.61	52.7	326.19	292.66

$$\Sigma X = 50.30$$
$$\Sigma X^2 = 269.61$$
$$\Sigma Y = 52.7$$
$$\Sigma Y^2 = 326.19$$
$$\Sigma XY = 292.66$$

It is very important to remember that correlation tells both the *strength* and the *direction* of an association. An r of $-.87$ describes a much more important and useful association than an r of .23. A negative association explains as much as a positive one. The *strength* of the association is indicated by the *size* of the r, not by its sign.

Hypothesis Tests about the Population Correlation

As our data are drawn from a sample—and a small one at that— we would be well advised to test the null hypothesis that the correlation between the two variables in the population is equal to 0. The symbol for the Pearson correlation in the population is ρ (*rho*); so we are testing the hypothesis that $\rho = 0$. We can then select a significance level and test to see if our sample r value is an unlikely sample value to obtain when there is no linear relationship between the variables in the population. Just as with hypothesis tests about means, if the true population correlation equals 0, we must look at the sampling distribution of the sample correlation r. That is, we must determine what sample r values will cause us to reject our null hpyothesis.

Fortunately, the procedure is easier than testing a hypothesis about means, because there is a ready-made table of the values of the correlation coefficient for various levels of significance and sample sizes (appendix table 5). To use the table, figure the degrees of freedom $(n - 2)$ and read the value for that df value under the chosen level of significance. For our example, $df = 8$. At the .05 level, we find in the table, a *minimum* correlation of .63 would be expected to reject the null hypothesis for 8 degrees of freedom.

That is, if the tabular value of r is less than the r we obtain, we can assume that our r is statistically significant, that it is far enough from a value of 0 for us to conclude that our assumed hypothesis is incorrect, and that there *is* a linear relation. This does *not* mean that there is any causal connection between the variables—only that the measure we find is not due to chance alone. *Other* factors (not measured) may be "causing" the relationship.

Regression Analysis

When, as in the preceding example, there is a strong (high) correlation between variables, we may wish to express the relationship in more precise terms. It is possible to construct a precise mathematical statement describing the hypothetical value of any Y (per capita circulation), given any value of X (per capita expenditure), and *regression analysis* is the classic statement of this kind. In effect, it enables us to predict a value of one variable based on a given value of another. It is based on the information in the correlated data. If these data are not representative, if the correlation is spurious or coincidental, no matter how "high," then any "regression" or prediction is valueless, if not misleading.

If we can assume that other branch libraries fit the pattern described by our sample, we may want to exploit our findings to predict the association between circulation and expenditure in other libraries. For example, we may ask, what per capita circulation can be expected in a branch that receives $8 per capita expenditure?

The formula $Y = a + bX$ will readily give us the precise per capita circulation that can be expected for any per capita expenditure within the range of our original X values ($3 − $7.20). If we try to predict per capita circulation based on per capita expenditures outside this range, we cannot be sure that the same sort of relationship holds. If we are willing to make the assumption that the relationship is the same outside the X range, and if it seems reasonable to assume so, we can. But we should keep this caveat in mind.

In this formula, Y is the dependent variable or the unknown per capita circulation for which we are "solving." $X = \$8$ is our independent variable, our projected per capita expenditure.

The formula

$$a = \frac{\Sigma Y - b\Sigma X}{n}$$

provides a figure that becomes a constant (intercept) in our regression analysis. The formula

$$b = \frac{n\Sigma XY - (\Sigma X)(\Sigma Y)}{n\Sigma X^2 - (\Sigma X)^2}$$

describes the *slope* of the regression line.

Results

As b is a factor in the formula for a, b must be obtained first.

$$a = \frac{\Sigma Y - b\Sigma X}{n}$$

$$b = \frac{n\Sigma XY - (\Sigma X)(\Sigma Y)}{n\Sigma X^2 - (\Sigma X)^2} = \frac{10\,(292.66) - (50.3)(52.7)}{10\,(269.61) - (50.30)^2}$$

$$= \frac{275.79}{166.01} = 1.66$$

$$a = \frac{52.7 - (1.66)\,50.30}{10} = \frac{52.7 - 83.50}{10} = \frac{-30.80}{10} = -3.08$$

We want to be able to use this formula

$$Y = -3.08 + 1.66X$$

to predict per capita circulation for cases for which we do not have per capita circulation data (Y) but *do* have per capita expenditure information (X).

Before we make any estimates of Y for any new cases, let us see how well our prediction equation fits our original data. We will use the given X values of each of the 10 cases and predict the Y value. If the predicted Y value ($\hat{Y}$) is very close to the real Y value, our regression equation is extremely useful, and our data have a strong linear relationship. If, however, the real Y values are very different from $\hat{Y}$, our linear model or regression equation is a poor fit to our data, and the linear association is a poor one.

For these data, we already know that the prediction will be nearly perfect, since the sample r value was calculated as .97. Nevertheless, for heuristic purposes we will demonstrate the relationship between Y and $\hat{Y}$. By placing each X value into the equation and completing the calculation, we arrive at the data in table 14. Notice that we have included library M, which has a per capita expenditure of $8 but no per capita circulation. Perhaps this information is missing because library M is a new library. We must assume that the linear relationship exists outside the range of our original X values in order to make a prediction for library M, but let us say that the assumption seems to be reasonable. We can now use our knowledge of regression analysis, using one independent variable (X) to predict what the value of Y would be for library M. Libraries A through J have data on both variables and therefore were used to calculate the coefficients in the formula

$$\hat{Y} = a + bX$$
$$\hat{Y} = -3.08 + 1.66 (\$8)$$

$\quad = 10.2$ volumes per capita expected circulation for the
$\quad\quad$ \$8 per capita expenditure of library M

At the end of the year it was found that library M had a real per capita circulation of 7.2. This was different from the predicted value of 10.2. So we can see that, generally, our predicted values will be a little (or a lot!) in error.

The stronger the relationship between the variables, the smaller the error. If the relationship is perfect ($r =$ either $+1$ or -1), there will be *no* error in prediction and we will always be exactly right. The mathematical model that we are trying to fit our data to in this example is a linear one. You can see, by examination of table 14, that there is a difference between the true Y value and the predicted Y value for each branch library. You would probably never want to calculate the predicted Y values for libraries A through J in reality, unless you

Table 14. **Data for 11 Cases on Variables** X **and** Y, **and Predicted Value of** Y
Calculated from Regression Equation

Case	Per Capita Expenditure Variable X	Per Capita Circulation Variable Y	$\hat{Y}$ Predicted per Capita Circulation
A	4.60	4.5	4.556
B	4.10	4.6	3.726
C	6.70	8.2	8.042
D	3.90	2.5	3.394
E	3.00	2.1	1.9
F	5.10	4.9	5.386
G	4.00	3.9	3.56
H	7.20	8.9	8.872
I	6.20	7.7	7.212
J	5.50	5.4	6.05
M	8.00		10.2

were interested in determining the amount of error in your prediction equation.

A number of different statistics can be calculated to determine exactly how good a regression equation is for prediction purposes. In addition, significant tests can be run on the coefficients in the equation, and in the equation itself. If you are interested in learning more about regression, consult the references, which provide a mathematical and extensive treatment of the topic.

As stated before, multiple linear regression analysis is generally the type of regression used in research. In that procedure we examine the relationships among many variables—specifically, many independent variables (X's) and one dependent variable (Y). In multiple regression, as in simple regression, the usual goal is to determine the *form* of the relationship among the variables of interest. If we can specify the type of relationship which exists mathematically, we can utilize this information to make predictions for individuals, measured on our X variables, for whom there are no data on the Y variable, our variable of interest.

Exercises

1. We are interested in the relationship between library use and grade-point average in English 151 among our undergraduate students. We have drawn a sample of 8 students, with the results below. Library use is measured by the number of books each student borrowed during the term for the English course for which we have the grade. Compute the correlation coefficient.

	Y	X
Student	No. of books borrowed	Grade in English 151
A	1	70
B	4	80
C	3	70
D	4	90
E	3	80
F	6	90
G	5	70
H	3	70

2. a. Determine the regression equation for predicting number of books borrowed, given your grade in English 151.
   b. If you have a grade of 75 in English 151, what would be your predicted number of books borrowed?

REFERENCES AND SUGGESTED READINGS

Blalock, Hubert M., Jr. *Social Statistics*. 2d ed. New York: McGraw-Hill, 1972, p. 361–383.

Ferguson, George A. *Statistical Analysis in Psychology and Education*. 4th ed. New York: McGraw-Hill, 1976.

Guildford, J. P. *Fundamental Statistics in Psychology and Education*. 5th ed. New York: McGraw-Hill, 1973.

Nie, Norman, Hadlai Hull, et al. *Statistical Package for the Social Sciences*. 2d ed. New York: McGraw-Hill, 1975.

Wonnacott, Thomas H., and Ronald J. Wonnacott. *Introductory Statistics*. 2d ed. New York: Wiley, 1972.

5
Nonparametric Tests and Measures

The many tests and measures labeled "nonparametric" may often be useful because they are simple to use and do not require assumptions about the shape of the population distribution of the variable or that the variable be measured at an interval scale of measurement. Many variables that are analyzed in library and information centers are not susceptible to a refined level of measurement but still are highly desirable for statistical treatment. The earlier examples on levels of measurement and those in this chapter will illustrate only a few such instances. Further, whenever one is not sure that an interval scale is really applicable, although it may seem to be, these procedures may be relied on. Nonparametric procedures do not require the assumption of an interval scale, but they may be applied to data that are so measured.

The temptation, in being "conservative," is to use nonparametric procedures when in doubt about the level of measurement. On the other hand, these procedures invite the possibility of avoiding rigorous conceptualizing and measurement. Some critics feel that measuring variables in such terms as "more than" or "less than," or "high" or "medium" or "low" (nominal scale), or rank-ordered as "first choice," "second choice" . . . "nth choice" (ordinal scale), leads to less exact measurement than is desirable or possible. Certainly it is important to measure things precisely in order to provide data that can generate the most precise or powerful conclusions, but, on the other hand, many variables still defy intervalscale measurement, and nonparametric procedures continue to fill a great need.

The term nonparametric may be misleading. It does *not* mean that these models have no parameters; it means that we are not required to make any explicit assumptions about the shape of the population distribution. It does not mean, however, that the tests are assumption-free!

Nominal Scale Tests and Measures

The Chi-Square Test of Independence

Chi-square (χ^2) is one of the most useful and, consequently, most common tools in statistics. It is often referred to as a "goodness of fit" statistic, because it is a measure of how much the observed frequencies differ from frequencies we would expect from the assumptions we made in our null hypothesis. The formula for χ^2 is often written:

$$\chi^2 = \sum \frac{(fo - fe)^2}{fe}$$

where fo is the observed frequency and fe is the expected frequency. That is, χ^2 is equal to the sum over all cells of the squared difference between the observed and expected frequencies divided by the expected frequency. The equation can be mathematically transformed into

$$\sum \frac{fo^2}{fe} - n$$

where n is the total number of cases or observations. As is true of all the techniques in this chapter, it does not require an interval, or even an ordinal level of measurement, although it may be applied to such data. It requires the following assumptions.

1. A random sample.
2. Each case is independent of every other case; that is, the value of the variable for one individual does not influence the value for another individual.
3. Each case can fall into one cell only.
4. There should be no expected frequency less than 1.
5. At least 80 percent of the expected frequencies should be greater than or equal to 5.

The chi-square statistic allows us to test for a relationship between the variables of interest in the population. We are not interested in discovering the form of the relationship at this point (i.e., linear), but whether it exists. As before, we assume that there is no relationship between the variables in the population. This is our null hypothesis: The two variables are operating independently of each other. We assume that this is true until our sample data are deviant enough for us to conclude otherwise.

Chi-square requires working with frequencies. If data are reported as percentages, they must be converted to frequencies. If 80 percent of a group of 150 users are female and 20 percent are male, it is necessary, for computation purposes, to convert the data to 120 females and 30 males. To test our null hypothesis, we must determine both the observed and the expected frequencies for each cell. The expected frequencies are those that we would expect to occur, if there is no relationship between the variables in the population. We then compare our observed with our expected frequencies. If the discrepancy is unusually large, we conclude that our assumption was in error and that a relationship *does* exist between the variables in the population.

Before we turn to a sample problem with given, observed data, it will be useful to see how the expected or probable frequencies are obtained —and consequently the logic of chi-square as an analysis of the difference between observed and probable frequencies.

Assume that we have just completed a random sample survey of 60 adults: 30 males and 30 females. Among the 60 people are 20 library users and 40 nonusers. We have two variables on which the one random sample of individuals has been measured: sex and library use. We can place each individual into one—and only one—cell of a "cross tabulation" table according to the value each individual has on each of the two variables. For the moment, consider the four cells as empty (we will not yet categorize each case). Let us suppose that only the "marginals" are known, that is, the number of males and females and the number of library users and nonusers.

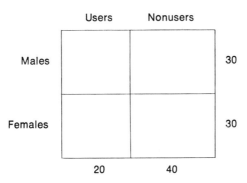

If we assume that there is no relationship between sex and library usage, males are no more likely to use the library than females. Let us try to fill in the expected cell values we would obtain if the two variables *are* independent.

Under this assumption, and on the basis of chance alone, what values would you expect to appear in each of the four vacant cells—how many male users, male nonusers, female users, female nonusers? Your logic should be something like this: Half of the total group (30 out of 60) is male; so I would expect half of the *users* to be male. Thus we compute expected frequencies based on the marginals, the row and column totals. This reasoning concludes that there are 10 male users, 10 female users, 20 male nonusers, and 20 female nonusers.

We can use a formula to determine the expected values, rather than reason, if we prefer. In this case the expected value of any cell equals its row marginal, times its column marginal, divided by the total sample size. The expected frequencies (*fe*) for each cell are

$$fe \text{ for cell a} = (30)(20) \div 60 = 10$$
$$fe \text{ for cell b} = (30)(40) \div 60 = 20$$
$$fe \text{ for cell c} = (30)(20) \div 60 = 10$$
$$fe \text{ for cell d} = (30)(40) \div 60 = 20$$

However, the actual "observations" from our survey show that, among our random sample of library users ($n = 20$), only 30 percent are male —an actual frequency of 6 male users. There are 24 male nonusers, 14 female users, and 16 female nonusers. Table 15 shows the data as found

Table 15. **Cross-Tabulation of Sex by Library Usage, with Expected Cell Frequencies in Upper Left-Hand Corner of Each Cell and Observed Frequencies in the Cell**

	Users			Nonusers		
Males	10	6	a	20	24	b	30
Females	10	14	c	20	16	d	30
		20			40		$n = 60$

in the survey, with the actual (observed) frequencies in the cells and with the expected probable frequencies in the corner of each cell for visual and computational comparison.

Our research question is as follows: Is this difference between actual and expected frequencies statistically significant? As with other tests of hypotheses, we may select a .05 significance level for this example.

The computation of chi-square is straightforward. From left to right, beginning in the upper left cell, the following values may be listed.

fo	fe	fo^2	fo^2 / fe
6	10	36	3.6
24	20	576	28.8
14	10	196	19.6
16	20	256	12.8
			64.8

$$\chi^2 = \Sigma fo^2 / fe - n = 64.8 - 60 = 4.8$$

A special table for the chi-square distribution, which is another special mathematical distribution, allows us to interpret the values from our computation. This table can be found as appendix table 4. We need only to determine the degrees of freedom and read the value in the table for the .05 level we have selected. The formula for degrees of freedom for chi-square is $(r - 1)(c - 1)$: the number of rows minus 1 times the number of columns minus 1. With two rows and two columns, there is 1 degree of freedom. Reading the chi-square table, we see that at the .05 level, with 1 degree of freedom if the variables are truly independent, we would expect a chi-square sample value no larger than 3.841.

Figure 21 illustrates the sampling distribution of the chi-square statistic when the variables in the population *are* independent (the null hypothesis is true). This distribution could be found by taking all possible samples (with $n = 60$ and the marginals given) and calculating the chi-square value for each sample. We would get a distribution of sample chi-square values, a sampling distribution in which some values would be likely and some would be unlikely to occur. If the variables are independent, we would expect the differences between fe and fo to be small and small chi-square statistics to result.

The shaded area in the tail represents our unlikely occurrences, which we have defined to be equal to 5 percent of the total area under the curve. The value 3.841 is on the edge of the shaded area. We will reject our hypothesis if our sample value falls to the right of 3.841. The computed value of 4.8 is statistically significant since it falls inside the shaded region of the figure and would be expected to occur less than 5 times in 100, if our hypothesis is true. Therefore, we reject the null hypothesis that there is no relationship between sex and library use, and conclude that these variables are indeed related.

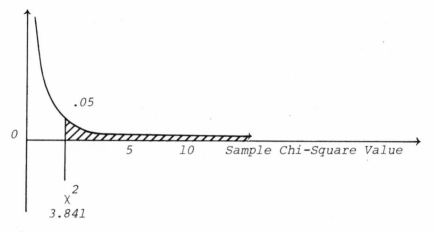

(Notice that the distribution consists
of only positive values.)

Figure 21. Chi-Square Distribution with 1° of Freedom Shown as a Sampling Distribution, with Shaded Region of Rejection Equaling 5% of Total Area

To repeat an earlier statement about significance levels in hypothesis testing, there is nothing sacred about any one given level, although .05 and .01 are very commonly used. Some prefer to report conclusions somewhat differently to give more information to the reader. For instance, the probability of obtaining a chi-square value of 4.8 with 1 degree of freedom falls between .05 and .01. Although we reject the null hypothesis at the .05 level, we would not at the .01 level.

The chi-square test can be misused. It is almost impossible not to obtain "significance" with very large samples. If we were to take a data base as large as the census of the United States, we would find nearly everything we might wish to turn out statistically significant by the use of chi-square. In other words, the user must beware of large numbers.

For example, a sample of 50 adults shows library use by sex, as indicated in the following cross-tabulation.

	Users	Nonusers	
Males	12.5 10	12.5 15	25
Females	12.5 15	12.5 10	25
	25	25	50

fo	fe	fo²	fo² / fe
10	12.5	100	8
15	12.5	225	18
15	12.5	225	18
10	12.5	100	8
			52

$$\chi^2 = 52 - 50 = 2$$

Not significant at the .05 level, 1 *df*.

Null hypothesis is not rejected; we have no basis for assuming that men and women differ in library use.

Initially, in this example, we probably would have noted that 40 percent of the men in our sample are library users $(10 / 25 = .40)$, 60 percent are nonusers $(15 / 25 = .60)$, and 60 percent of women use the library $(15 / 25 = .60)$ whereas 40 percent $(10 / 25 = .40)$ do not—a finding that would invite a test of the hypothesis that sex is related to library use. Now consider retaining the same percentages but increasing the sample to 5,000.

	Users	Nonusers	
Males	1,250 1,000	1,250 1,500	2,500
Females	1,250 1,500	1,250 1,000	2,500
	2,500	2,500	5,000

fo	fe	fo²	fo² / fe
1,000	1,250	1,000,000	800
1,500	1,250	2,250,000	1,800
1,500	1,250	2,250,000	1,800
1,000	1,250	1,000,000	800
			5,200

$$\chi^2 = 5,200 - 5,000 = 200$$

Significant at the .001 level. We reject the null hypothesis and conclude that there is a relationship between sex and library use.

Obviously, the only thing that has changed is the size of the sample; the *proportions* of men and women using the library are the same. The inherent problem is that chi-square, like other hypothesis tests, is susceptible to large numbers! One's faith in the procedure might be shaken at this. On the contrary, however, remember that, first of all, summarizing by percentages or proportions tells us a great deal and, second, that chi-square must be used cautiously with large samples.

The use of chi-square can be extended to tables larger than two rows and two columns, and even more than two variables. A few years ago, Maurice Marchant analyzed the faculty's use of the public library in Ann Arbor, Michigan.[1] Among the results, he found that 42 of 113 psychologists, 17 of 68 biologists, 33 of 203 engineers, and 20 of 78 English professors were users of the public library. The percentages were 37 in psychology, 25 in biology, 16 in engineering, and 26 in English. Were these results statistically significant? Marchant says they were, and we may check on the validity of his claim by setting up a table, as follows, and applying the chi-square procedure.

	Psychology	Biology	Engineers	English	
Users	27 42	16 17	49 33	19 20	112
Nonusers	86 71	52 51	154 170	59 58	350
	113	68	203	78	462

1. Maurice Marchant, "Faculty as Public Library Patrons," *Wilson Library Bulletin* (January 1969), p. 445–446.

The numbers in the center of each cell represent those who are found to be users or nonusers. The numbers in the upper left corner of the cells represent those we would expect to find by chance. The expected values for psychology nonusers and English nonusers will be calculated to refresh our memory:

$$fe = \frac{(350)(113)}{462} \text{ (Psychology nonusers)}$$

$$fe = \frac{(350)(78)}{462} \text{ (English nonusers)}$$

We will now find the chi-square statistic for a random sample of cases.

fo	fe	fo²	fo² / fe
42	27	1,764	65.3
17	16	289	18.1
33	49	1,089	22.2
20	19	400	21.1
71	86	5,041	58.6
51	52	2,601	50.0
170	154	28,900	187.7
58	59	3,364	57.0
			480.0 − 462 = 18 = χ^2

With four columns and two rows, we apply the formula for degrees of freedom $(r-1)$ $(c-1)$ and find 3 degrees of freedom. The chi-square table shows that our chi-square of 18 is statistically significant at the .001 level. There seems to be a relationship between department and library usage.

Strength of Association

There are many measures that will tell us the strength of the association between the two variables, given that they are significantly related. Those under discussion here are all functions of chi-square and are accordingly limited. Phi-square, ϕ^2, is the simplest of the group and requires merely dividing the chi-square value by n, that is, $\phi^2 = \chi^2 / n$. In all the examples noted above, where the proportions were retained but the size of the sample increased, ϕ^2 remains the same. In effect, it provides a kind of control over sample size affecting significance. In a 2×2 table, ϕ^2 may reach a value of 1. The following cross-tabulation shows a "perfect" relationship.

12.5		12.5		
	25		0	25
12.5		12.5		
	0		25	25
	25		25	50

The chi-square value will work out to be 50. $\phi^2 = \chi^2 / n = 1.0$: a "perfect" relationship.

Other measures include

$$T^2 = \frac{\phi^2}{\sqrt{(r-1)\,(c-1)}}$$

$$V^2 = \frac{\phi^2}{\text{Min}\,[(r-1)\,(c-1)]}$$

(Use the number of rows *or* columns, whichever is smaller.)

In 2 $\times$ 2 analyses, ϕ^2, T^2, and V^2 are equal to each other.

The Contingency Coefficient

C is also known as *Pearson's C* or *Pearson's contingency coefficient.* It is also based on chi-square, and is probably the most widely used of this group of measures. *C* has a value of 0 when there is no association. It does not have unity as a maximum value when the variables are completely related or dependent on each other. In a 2 $\times$ 2 case, the maximum value of *C* is .707; in a 3 $\times$ 3 table, .816. Indeed, when the numbers of rows and columns are equal to each other, the upper limit of *C* is $\sqrt{\dfrac{\text{columns} - 1}{\text{columns}}}$.

As with so many other measures, *C* is widely used because it requires only a nominal scale of measurement—and is easy to calculate. Contingency coefficients can be compared with each other when they are based on tables of the same size (i.e., tables that have the same number of rows and columns). They cannot be compared directly to other measures of association that are designed for interval and ordinal scales, such as Pearson's *r* and Spearman's r_s.

The formula for the contingency coefficient is

$$C = \sqrt{\frac{\chi^2}{\chi^2 + n}}$$

A report for two variables of a chi-square value of 15, with an n of 50, would produce a C of .48. This indicates a high degree of association, given that the maximum value possible for C in a 2×2 table is .707. To our earlier example of chi-square, where we found a significant relationship between sex and library use, we apply the contingency coefficient, and the C value is .20. This is a moderately low degree of association, given .707 as maximum possibility.

Ordinal Scale Tests and Measures

The Wald–Wolfowitz Runs Test

This procedure can be used when we want to test the null hypothesis that two samples are from the same population. It requires that we have independent random samples and at least an ordinal scale of measurement, but we do not need to be concerned about the population(s) being normal. In effect, the runs test is a test of the form and overall dispersion of the populations. It is a test of central tendency and variability that, at the same time, does not pinpoint differences in a single measure such as central tendency. The null hypothesis is that the two samples come from populations with the same distribution, and we assume that this is true until our sample data indicate otherwise.

Let us assume, for a sample problem, that we have asked students to *rank* a random sample of 16 reference books according to their usefulness to the students. The books are to be ranked by respondents, using a score of 1 for the title considered most useful and 16 for the title deemed least useful. In this sample collection of 16 titles are 8 titles that were selected by faculty members and 8 selected by the library staff. Our question is: Is there a statistically significant difference in the usefulness of reference books selected by faculty and by librarians? Our null hypothesis is that there is no statistically significant difference between the faculty's and the librarians' selections.

Performing the test involves little if any computation. If $n_1 + n_2$ are equal to or less than 20, use appendix table 5. For n's larger than 20, the sampling distribution of R is nearly normal, and the following formulas may be used.

$$\text{Mean} = \mu_R = \frac{2n_1n_2}{n_1 + n_2} + 1$$

$$\text{Standard deviation} = \sigma_R = \sqrt{\frac{2n_1n_2\,(2n_1n_2 - n_1 - n_2)}{(n_1 + n_2)^2\,(n_1 + n_2 - 1)}}$$

In our sample problem, the total n is 16, which is less than 20; so we may use the table of critical values of R in the runs test at the .05 level of significance. To do so, first arrange the books in order of their usefulness, underlining those titles that were selected by librarians and overlining those selected by faculty. Let us assume that we obtain the following results:

$$\underline{1}\ \overline{2\ 3}\ \underline{4\ 5\ 6\ 7}\ \overline{8}\ \underline{9}\ \overline{10}\ \underline{11\ 12}\ \overline{13\ 14\ 15\ 16}$$

At this point we have simply delineated the titles selected by faculty and those by librarians. The title ranked most useful (1) was selected by the library staff; the titles ranked second and third were among those selected by faculty; etc. We obtain the number of runs simply by counting the number of lines, that is, 1 line is equal to a run. There are 8 runs. Using the table, we read down the n column to 8, the size of our n_1, and across the n row to 8, the size of our n_2. There we find that, with samples of this size, we would expect to find *5 or fewer* runs at the .05 level of significance. Our test result of $R = 8$ runs, being larger than that predicted by the distribution in the table, is *not* significant. We do not reject the null hypothesis at the .05 level.

This example is really too small to require using the formulas, when $n_1 + n_2$ exceeds 20, but we can use it for demonstration purposes.

$$(\mu_R)\ \text{Mean} = \frac{2\,(8 \times 8)}{(8 + 8)} + 1 = \frac{128}{16} + 1 = 9$$

$$(\sigma_R)\ \text{Standard deviation} = \sqrt{\frac{2\,(8 \times 8)\,(2\,(8 \times 8) - 8 - 8)}{(8 + 8)^2\,(8 + 8 - 1)}} = 1.93$$

$$Z = \frac{R - \mu_R}{\sigma_R} = \frac{8 - 9}{1.93} = -.5$$

Using the table of areas under the normal curve, we find that Z, at the .05 level, is greater than 1.96. We do not reject the null hypothesis. (We would reject it only if our Z were equal to or less than the values at 1.96.)

Consider what the runs would look like if there were perfect significance. Something like this might result:

$$\overline{1\ 2\ 3\ 4\ 5\ 6\ 7\ 8}\ \overline{9\ 10\ 11\ 12\ 13\ 14\ 15\ 16}$$

We would have obtained only two runs. (All titles selected by librarians
outrank all those selected by faculty.) The greater the number of runs,
the more likely that there is random nonassociation between the two
samples.

Spearman's Rank-Order Correlation, r_s

This procedure allows us to measure the strength of association between ordinally scaled variables. r_s has a maximum value of $+1$ when
ranks are in perfect agreement and -1 when they are in perfect disagreement.

As an example, assume that we have asked the undergraduate and
graduate students at our school to rank-order their preferences for 5
uses of leisure time: books, television, radio, magazines, and movies.
r_s will tell us how closely the graduate students and undergraduate students agree on the preference of these activities in their leisure time.
Assume that the results are as follows.

	Undergraduates	Graduates	Di	Di^2
Television	1	2	-1	1
Movies	2	3	-1	1
Books	3	1	2	4
Magazines	4	4	0	0
Radio	5	5	0	0
				6

Undergraduates ranked television as their first preference for leisure-
time activity, graduate students ranked television as their second choice,
etc. The Di column indicates the difference between the two groups'
rankings. The strength of the association, the strength of the agreement
between the rankings, is computed by the formula

$$r_s = 1 - \frac{6\Sigma(Di)^2}{n(n^2-1)} = 1 - \frac{6\,(6)}{5\,(25-1)} = .70$$

where n is equal to the total number of variables being ranked. The
moderately high association of .70 is not surprising, as we can scan the

table of this simple example and see that graduates and undergraduates are generally agreed on how they prefer to use their leisure time; indeed, on two items they are in perfect agreement.

To further illuminate how r_s may be interpreted, let us see what would happen in the case of a "perfect disagreement," an r_s value of -1.

	Graduates	Undergraduates	Di	Di²
Activity A	1	5	−4	16
Activity B	2	4	−2	4
Activity C	3	3	0	0
Activity D	4	2	2	4
Activity E	5	1	4	16
				40

$$r_s = 1 - \frac{6 \ (40)}{5(25 - 1)} = 1 - \frac{240}{120} = 1 - 2 = -1$$

In this case, graduates and undergraduates have ranked the total list of activities in exactly the reverse order. r_s shows this in its maximum negative value of -1. This does not tell us the percentage of agreement between the two groups—it *does* tell us the degree to which the two rank-ordered *sets* of activities are in agreement. In this case, there is perfect disagreement between the two *sets*, although there is agreement on one activity in the sets (activity C).

Use of r_s with Interval Scale Data

Sometimes we have reason to doubt the precision of measurements that are reported in interval scale. The actual size of large library collections, for instance, is often open to question. Knowing the difficulties in getting an accurate count of such collections, we may wonder about the "real" difference between a library reporting 2,332,650 volumes and another reporting 2,340,960. Furthermore, if we were correlating the size of collection with another large variable, such as institutional budget, the calculations would become quite difficult. The use of r_s may be helpful in such cases.

For illustration, we may measure the association between the size of population served by a library and the size of the library's collection. We may ask: Does the library's collection increase as the population it serves increases? Following are data from 10 libraries in a regional system.

Library	Total Population	Rank	Total Collection	Rank	Di	Di²
A	38,731	10	55,756	10	0	0
B	66,281	8	83,866	7	1	1
C	125,916	2	109,527	5	−3	9
D	103,624	5	106,840	6	−1	1
E	46,248	9	83,404	8	1	1
F	181,097	1	282,081	1	0	0
G	102,676	6	79,861	9	−3	9
H	79,950	7	147,568	3	4	16
I	107,930	4	151,176	2	2	4
J	121,692	3	114,098	4	−1	1
						42

$$r_s = 1 - \frac{6\,(Di^2)}{n\,(n-1)} = 1 - \frac{6\,(42)}{10\,(99)} = 1 - \frac{252}{990} = 1 - .25 = .75$$

The computation of Pearson's r, without mechanical assistance, would be difficult indeed, requiring the manipulation of figures running as large as 12 digits. The results would yield an r of .80, not substantially higher than the easily calculated r_s. When time and equipment for calculating Pearson's r are not available, r_s may be used as a reasonable estimate.

Kendall's *Tau*

This procedure also is used to obtain a measure of association between ordinal scales. It requires that we have at least an ordinal level of measurement. *Tau*, like r_s, ranges in value from $+1$. to -1. Applying it to the same example we used for r_s, we may compare the two measures.

	a TV	b Movies	c Books	d Magazines	e Radio
Undergraduates	1	2	3	4	5
Graduates	2	3	1	4	5

Our question is: What pairs of activities are ranked in the same direction or order by graduates and undergraduates alike? The activities are labeled a, b, c, d, and e to facilitate examination of the activities in pairs. To determine *tau*, we must order the ranks of one set (undergraduates) from lowest to highest. It is not necessary to do this when we calculate r_s. We assign a plus-1 value to each pair in the same order and a minus 1 to each pair in opposite order. For example, the pair a, b (television and movies) is ranked in the same order. That is, graduates *and* undergraduates rank television higher than movies. All possible pairs in this example are measured as follows to obtain the statistic, S.

$$
\begin{array}{llll}
a, b + a, c + a, d + a, e & = +1 -1 +1 +1 & = +2 \\
b, c + b, d + b, e & = -1 +1 +1 & = +1 \\
c, d + c, e & = +1 +1 & = +2 \\
d, e & = +1 & = \underline{+1} \\
& & S = \quad 6
\end{array}
$$

$$
tau = \frac{S}{n\,(n-1)/2} = \frac{6}{(5)\,(4)/2} = .60
$$

Our *tau* of .60 seems to indicate a lower strength of association than did the r_s of .70. This is very often the case. By squaring the differences between ranks, r_s tends to give relatively more weight to extreme differences, while *tau* gives equal weight to all differences.

Let's consider a second example of Kendall's *tau*. Before a management-by-objectives training program, the following objectives were ranked by administration and staff of the library: higher salaries, reduced staff, higher rate of production, responsibility and more clearly linked authority, improved fringe benefits, and stronger middle management.

	a	b	c	d	e	f
			Rank of Objectives			
	Higher Produc- tivity	Middle Manage- ment	Responsi- bility/ Authority	Smaller Staff	Fringe Benefits	Salaries
Administration	1	2	3	4	5	6
Staff	5	4	2	6	3	1

$$
\begin{array}{llll}
a, b + a, c + a, d + a, e + a, f & = -1 -1 +1 -1 -1 & = -3 \\
b, c + b, d + b, e + b, f & = -1 +1 -1 -1 & = -2 \\
c, d + c, e + c, f & = +1 +1 -1 & = +1 \\
d, e + d, f & = -1 -1 & = -2 \\
e, f & = -1 & = \underline{-1} \\
& & S = \overline{-7}
\end{array}
$$

$$
tau = \frac{S}{n\,(n-1)/2} = \frac{-7}{(6)\,(5)/2} = -.47
$$

Our calculation of *tau* indicates a moderate inverse or negative association: administrators and staff tend to rank objectives in opposite ways.

When the number of objects to be ranked is large, the calculation of *tau* (in the fashion presented here) can be unwieldy. See Glass and Stanley (at the end of this chapter) for other approaches to the calculation of *tau*.

An important problem, which we have avoided discussing in this section, is that of tied ranks. In other words, what do we do when two objects which are to be ranked are given the same ranking? In our previous example, what would happen if administration could not distinguish between responsibility and smaller staff in terms of rank of objectives and felt that both are of equal importance? When tied ranks occur in the calculation of either *tau* or r_s, the coefficients are affected and certain adjustments must be made to correct for the tied occurrences.

If you are interested in making inferences from an r_s, or *tau* sample statistic to a population parameter, you will have to employ hypothesis-testing procedures. We have chosen not to deal with these advanced topics in our introductory discussion and have, instead, provided references at the end of this chapter (Hays and Siegel) which will assist you in handling these situations.

The procedures in this chapter require few assumptions by the researcher. We do not have to be concerned about the applicability of our methods, if we are not aware of the shape of the distribution of our variable(s) in the population or if we have less than an interval level of measurement. Of course, whenever parametric measures and procedures are appropriate, we should attempt to employ them. Parametric methods are more powerful than nonparametric methods.

In other words, if there is a difference between our groups, or if there is a relationship between our variables, we are more likely to detect it with parametric procedures. Nevertheless, in many instances we do not feel that their use is warranted, due to lack of information or poor measurement instruments. It is in situations such as these that nonparametrics are useful.

We feel these techniques are most useful to those in library-related professions. They are but a small sample of techniques which fall under the classification of nonparametric statistics. We hope that we have given a basic understanding of their usefulness and we encourage readers to utilize them in their professional undertakings and to investigate them more fully if they find the need.

Exercise

1. a. Marchant's study, cited earlier, reports further findings about faculty use of the public library. He found that 53 of 235 full professors, 33 of 130 associate professors, and 26 of 97 assistant professors use the public library. Is faculty rank independent of usage? Use the chi-square test to answer this question.
   b. What is the value of C?

REFERENCES AND SUGGESTED READINGS

Babbie, Earl R. *The Practice of Social Research.* Belmont, Calif.: Wadsworth, 1975.

Blalock, Hubert M., Jr. *Social Statistics.* 2d ed. New York: McGraw-Hill, 1972, p. 249–254, 275–287, 291–302, 418–426.

Glass, Gene, and Julian C. Stanley. *Statistical Methods in Education and Psychology.* Englewood Cliffs, N.J.: Prentice-Hall, 1970.

Hays, W. L. *Statistics for the Social Sciences.* 2d ed. New York: Holt, Rinehart and Winston, 1973.

Kerlinger, Fred N. *Foundations of Behavioral Research.* 2d ed. New York: Holt, Rinehart and Winston, 1973.

Siegel, Sidney. *Non-Parametric Statistics.* New York: McGraw-Hill, 1956.

Appendix Tables

1. Table of Random Digits

2. Areas under the Normal Curve

3. Distribution of t

4. Distribution of χ^2

5. Values of the Correlation Coefficient

6. Critical Values of R in the Runs Test

1. TABLE OF RANDOM DIGITS

94015	46874	32444	48277	59820	96163	64654	25843	41145	42820
74108	88222	88570	74015	25704	91035	01755	14750	48968	38603
62880	87873	95160	59221	22304	90314	72877	17334	39283	04149
11748	12102	80580	41867	17710	59621	06554	07850	73950	79552
17944	05600	60478	03343	25852	58905	57216	39618	49856	99326
66067	42792	95043	52680	46780	56487	09971	59481	37006	22186
54244	91030	45547	70818	59849	96169	61459	21647	87417	17198
30945	57589	31732	57260	47670	07654	46376	25366	94746	49580
69170	37403	86995	90307	94304	71803	26825	05511	12459	91314
08345	88975	35841	85771	08105	59987	87112	21476	14713	71181
27767	43584	85301	88977	29490	69714	73035	41207	74699	09310
13025	14338	54066	15243	47724	66733	47431	43905	31048	56699
80217	36292	98525	24335	24432	24896	43277	58874	11466	16082
10875	62004	90391	61105	57411	06368	53856	30743	08670	84741
54127	57326	26629	19087	24472	88779	30540	27886	61732	75454
60311	42824	37301	42678	45990	43242	17374	52003	70707	70214
49739	71484	92003	98086	76668	73209	59202	11973	02902	33250
78626	51594	16453	94614	39014	97066	83012	09832	25571	77628
66692	13986	99837	00582	81232	44987	09504	96412	90193	79568
44071	28091	07362	97703	76447	42537	98524	97831	65704	09514
41468	85149	49554	17994	14924	39650	95294	00556	70481	06905
94559	37559	49678	53119	70312	05682	66986	34099	74474	20740
41615	70360	64114	58660	90850	64618	80620	51790	11436	38072
50273	93113	41794	86861	24781	89683	55411	85667	77535	99892
41396	80504	90670	08289	40902	05069	95083	06783	28102	57816

25807	24260	71529	78920	72682	07385	90726	57166	98884	08583
06170	97965	88302	98041	21443	41808	68984	83620	89747	93882
60808	54444	74412	81105	01176	28838	36421	16489	18059	51061
80940	44893	10408	36222	80582	71944	92638	40333	67054	16067
19516	90120	46759	71643	13177	55292	21036	82808	77501	97427
49386	54480	23604	23554	21785	41101	91178	10174	29420	90438
06312	88940	15995	69321	47458	64809	98189	81851	29651	84215
60942	00307	11897	92674	40405	68032	96717	54244	10701	41393
92329	98932	78284	46347	71209	92061	39448	93136	25722	08564
77936	63574	31384	51924	85561	29671	58137	17820	22751	36518
38101	77756	11657	13897	95889	57067	47648	13885	70669	93406
39641	69457	91339	22502	92613	89719	11947	56203	19324	20504
84054	40455	99396	63680	67667	60631	69181	96845	38525	11600
47468	03577	57649	63266	24700	71594	14004	23153	69249	05747
43321	31370	28977	23896	76479	68562	62342	07589	08899	05985
64281	61826	18555	64937	13173	33365	78851	16499	87064	13075
66847	70495	32350	02985	86716	38746	26313	77463	55387	72681
72461	33230	21529	53424	92581	02262	78438	66276	18396	73538
21032	91050	13058	16218	12470	56500	15292	76139	59526	52113
95362	67011	06651	16136	01016	00857	55018	56374	35824	71708
49712	97380	10404	55452	34030	60726	75211	10271	36633	68424
58275	61764	97586	54716	50259	46345	87195	46092	26787	60939
89514	11788	68224	23417	73959	76145	30342	40277	11049	72049
15472	50669	48139	36732	46874	37088	73465	09819	58869	35220
12120	86124	51247	44302	60883	52109	21437	36786	49226	77837

SOURCE: The Rand Corporation, *A Million Random Digits with 100,000 Normal Deviates* (Glencoe, Ill.: Free Press, 1955), p. 4. Reprinted with permission of the Rand Corporation.

2. AREAS UNDER THE NORMAL CURVE

x/σ	.00	.01	.02	.03	.04	.05	.06	.07	.08	.09
0.0	0000	0040	0080	0120	0159	0199	0239	0279	0319	0359
0.1	0398	0438	0478	0517	0557	0596	0636	0675	0714	0753
0.2	0793	0832	0871	0910	0948	0987	1026	1064	1103	1141
0.3	1179	1217	1255	1293	1331	1368	1406	1443	1480	1517
0.4	1554	1591	1628	1664	1700	1736	1772	1808	1844	1879
0.5	1915	1950	1985	2019	2054	2088	2123	2157	2190	2224
0.6	2257	2291	2324	2357	2389	2422	2454	2486	2518	2549
0.7	2580	2612	2642	2673	2704	2734	2764	2794	2823	2852
0.8	2881	2910	2939	2967	2995	3023	3051	3078	3106	3133
0.9	3159	3186	3212	3238	3264	3289	3315	3340	3365	3389
1.0	3413	3438	3461	3485	3508	3531	3554	3577	3599	3621
1.1	3643	3665	3686	3718	3729	3749	3770	3790	3810	3830
1.2	3849	3869	3888	3907	3925	3944	3962	3980	3997	4015
1.3	4032	4049	4066	4083	4099	4115	4131	4147	4162	4177
1.4	4192	4207	4222	4236	4251	4265	4279	4292	4306	4319
1.5	4332	4345	4357	4370	4382	4394	4406	4418	4430	4441
1.6	4452	4463	4474	4485	4495	4505	4515	4525	4535	4545
1.7	4554	4564	4573	4582	4591	4599	4608	4616	4625	4633
1.8	4641	4649	4656	4664	4671	4678	4686	4693	4699	4706
1.9	4713	4719	4726	4832	4738	4744	4750	4758	4762	4767
2.0	4773	4778	4783	4788	4793	4798	4803	4808	4812	4817
2.1	4821	4826	4830	4834	4838	4842	4846	4850	4854	4857
2.2	4861	4865	4868	4871	4875	4878	4881	4884	4887	4890
2.3	4893	4896	4898	4901	4904	4906	4909	4911	4913	4916
2.4	4918	4920	4922	4925	4927	4929	4931	4932	4934	4936

x/σ	.00	.01	.02	.03	.04	.05	.06	.07	.08	.09
2.5	4938	4940	4941	4943	4945	4946	4948	4949	4951	4952
2.6	4953	4955	4956	4957	4959	4960	4961	4962	4963	4964
2.7	4965	4966	4967	4968	4969	4970	4971	4972	4973	4974
2.8	4974	4975	4976	4977	4977	4978	4879	4980	4980	4981
2.9	4981	4983	4983	4984	4984	4984	4985	4985	4986	4986

x/σ	.00	.01	.02	.03	.04	.05	.06	.07	.08	.09
3.0	4986.5	4987	4987	4988	4988	4988	4989	4989	4989	4990
3.1	4990.3	4991	4991	4991	4992	4992	4992	4992	4993	4993
3.2	4993.129									
3.3	4995.166									
3.4	4996.631									
3.5	4997.674									
3.6	4998.409									
3.7	4998.922									
3.8	4999.277									
3.9	4999.519									
4.0	4999.683									
4.5	4999.966									
5.0	4999.997133									

Fractional parts of the total number (10,000) under the normal probability curve, corresponding to the distances on the baseline between the mean and successive points of division laid off from the mean. Distances are measured in units of the standard deviation, σ. To illustrate, the table is read as follows: between the mean ordinate, y, and any ordinate erected at a distance from it of, say, 8σ (i.e., $\frac{X}{\sigma} = .8$) is included in 28.81 percent of the entire area.

SOURCE: Harold O. Rugg, *Statistical Methods Applied to Education* (New York: Houghton Mifflin, 1917), p. 389–90. Copyright © 1917 by Houghton Mifflin Company. Used with permission.

3. DISTRIBUTION OF t

df	Level of significance for one-tailed test					
	.10	.05	.025	.01	.005	.0005
	Level of significance for two-tailed test					
	.20	.10	.05	.02	.01	.001
1	3.078	6.314	12.706	31.821	63.657	636.619
2	1.886	2.920	4.303	6.965	9.925	31.598
3	1.638	2.353	3.182	4.541	5.841	12.941
4	1.533	2.132	2.776	3.747	4.604	8.610
5	1.476	2.015	2.571	3.365	4.032	6.859
6	1.440	1.943	2.447	3.143	3.707	5.959
7	1.415	1.895	2.365	2.998	3.499	5.405
8	1.397	1.860	2.306	2.896	3.355	5.041
9	1.383	1.833	2.262	2.821	3.250	4.781
10	1.372	1.812	2.228	2.764	3.169	4.587
11	1.363	1.796	2.201	2.718	3.106	4.437
12	1.356	1.782	2.179	2.681	3.055	4.318
13	1.350	1.771	2.160	2.650	3.012	4.221
14	1.345	1.761	2.145	2.624	2.977	4.140
15	1.341	1.753	2.131	2.602	2.947	4.073
16	1.337	1.746	2.120	2.583	2.921	4.015
17	1.333	1.740	2.110	2.567	2.898	3.965
18	1.330	1.734	2.101	2.552	2.878	3.922
19	1.328	1.729	2.093	2.539	2.861	3.883
20	1.325	1.725	2.086	2.528	2.845	3.850

21	1.323	1.721	2.080	2.518	2.831	3.819
22	1.321	1.717	2.074	2.508	2.819	3.792
23	1.319	1.714	2.069	2.500	2.807	3.767
24	1.318	1.711	2.064	2.492	2.797	3.745
25	1.316	1.708	2.060	2.485	2.787	3.725
26	1.315	1.706	2.056	2.479	2.779	3.707
27	1.314	1.703	2.052	2.473	2.771	3.690
28	1.313	1.701	2.048	2.467	2.763	3.674
29	1.311	1.699	2.045	2.462	2.756	3.659
30	1.310	1.697	2.042	2.457	2.750	3.646
40	1.303	1.684	2.021	2.423	2.704	3.551
60	1.296	1.671	2.000	2.390	2.660	3.460
120	1.289	1.658	1.980	2.358	2.617	3.373
∞	1.282	1.645	1.960	2.326	2.576	3.291

SOURCE: Table III of Fisher and Yates, *Statistical Tables for Biological, Agricultural, and Medical Research*, published by Longman Group Ltd., London (previously published by Oliver and Boyd, Edinburgh). Used with permission of authors and publishers.

4. DISTRIBUTION OF χ^2

df	.99	.98	.95	.90	.80	.70	.50	.30	.20	.10	.05	.02	.01	.001
1	.0^{1}157	.0^{4}628	.00393	.0158	.0642	.148	.455	1.074	1.642	2.706	3.841	5.412	6.635	10.827
2	.0201	.0404	.103	.211	.446	.713	1.386	2.408	3.219	4.605	5.991	7.824	9.210	13.815
3	.115	.185	.352	.584	1.005	1.424	2.366	3.665	4.642	6.251	7.815	9.837	11.341	16.268
4	.297	.429	.711	1.064	1.649	2.195	3.357	4.878	5.989	7.779	9.488	11.668	13.277	18.465
5	.554	.752	1.145	1.610	2.343	3.000	4.351	6.064	7.289	9.236	11.070	13.388	15.086	20.517
6	.872	1.134	1.635	2.204	3.070	3.828	5.348	7.231	8.558	10.645	12.592	15.033	16.812	22.457
7	1.239	1.564	2.167	2.833	3.822	4.671	6.346	8.383	9.803	12.017	14.067	16.622	18.475	24.322
8	1.646	2.032	2.733	3.490	4.594	5.527	7.344	9.524	11.030	13.362	15.507	18.168	20.090	26.125
9	2.088	2.532	3.325	4.168	5.380	6.393	8.343	10.656	12.242	14.684	16.919	19.679	21.666	27.877
10	2.558	3.059	3.940	4.865	6.179	7.267	9.342	11.781	13.442	15.987	18.307	21.161	23.209	29.588
11	3.053	3.609	4.575	5.578	6.989	8.148	10.341	12.899	14.631	17.275	19.675	22.618	24.725	31.264
12	3.571	4.178	5.226	6.304	7.807	9.034	11.340	14.011	15.812	18.549	21.026	24.054	26.217	32.909
13	4.107	4.765	5.892	7.042	8.634	9.926	12.340	15.119	16.985	19.812	22.362	25.472	27.688	34.528
14	4.660	5.368	6.571	7.790	9.467	10.821	13.339	16.222	18.151	21.064	23.685	26.873	29.141	36.123
15	5.229	5.985	7.261	8.547	10.307	11.721	14.339	17.322	19.311	22.307	24.996	28.259	30.578	37.697
16	5.812	6.614	7.962	9.312	11.152	12.624	15.338	18.418	20.465	23.542	26.296	29.633	32.000	39.252
17	6.408	7.255	8.672	10.085	12.002	13.531	16.338	19.511	21.615	24.769	27.587	30.995	33.409	40.790
18	7.015	7.906	9.390	10.865	12.857	14.440	17.338	20.601	22.760	25.989	28.869	32.346	34.805	42.312
19	7.633	8.567	10.117	11.651	13.716	15.352	18.338	21.689	23.900	27.204	30.144	33.687	36.191	43.820
20	8.260	9.237	10.851	12.443	14.578	16.266	19.337	22.775	25.038	28.412	31.410	35.020	37.566	45.315
21	8.897	9.915	11.591	13.240	15.445	17.182	20.337	23.858	26.171	29.615	32.671	36.343	38.932	46.797
22	9.542	10.600	12.338	14.041	16.314	18.101	21.337	24.939	27.301	30.813	33.924	37.659	40.289	48.268
23	10.196	11.293	13.091	14.848	17.187	19.021	22.337	26.018	28.429	32.007	35.172	38.968	41.638	49.728
24	10.856	11.992	13.848	15.659	18.062	19.943	23.337	27.096	29.553	33.196	36.415	40.270	42.980	51.179
25	11.524	12.697	14.611	16.473	18.940	20.867	24.337	28.172	30.675	34.382	37.652	41.566	44.314	52.620

26	12.198	13.409	15.379	17.292	19.820	21.792	25.336	29.246	31.795	35.563	38.885	42.856	45.642	54.052
27	12.879	14.125	16.151	18.114	20.703	22.719	26.336	30.319	32.912	36.741	40.113	44.140	46.963	55.476
28	13.565	14.847	16.928	18.939	21.588	23.647	27.336	31.391	34.027	37.916	41.337	45.419	48.278	56.893
29	14.256	15.574	17.708	19.768	22.475	24.577	28.336	32.461	35.139	39.087	42.557	46.693	49.588	58.302
30	14.953	16.306	18.493	20.599	23.364	25.508	29.336	33.530	36.250	40.256	43.773	47.962	50.892	59.703

NOTE: For larger values of df, the expression $\sqrt{2\chi^2} - \sqrt{2df - 1}$ may be used as a normal deviate with unit variance, remembering that the probability for χ^2 corresponds with that of a single tail of the normal curve.

SOURCE: Table IV of Fisher and Yates, *Statistical Tables for Biological, Agricultural, and Medical Research*, published by Longman Group, Ltd., London (previously published by Oliver and Boyd, Edinburgh). Used with permission of the authors and publishers.

5. VALUES OF THE CORRELATION COEFFICIENT FOR DIFFERENT LEVELS OF SIGNIFICANCE

NOTE: The probabilities given are for a two-tailed test of significance, that is with the sign of r ignored. For a one-tailed test of significance, the tabled probabilities should be halved.

df	P = .10	.05	.02	.01
1	.988	.997	.9995	.9999
2	.900	.950	.980	.990
3	.805	.878	.934	.959
4	.729	.811	.882	.917
5	.669	.754	.833	.874
6	.622	.707	.789	.834
7	.582	.666	.750	.798
8	.549	.632	.716	.765
9	.521	.602	.685	.735
10	.497	.576	.658	.708
11	.476	.553	.634	.684
12	.458	.532	.612	.661
13	.441	.514	.592	.641
14	.426	.497	.574	.623
15	.412	.482	.558	.606
16	.400	.468	.542	.590
17	.389	.456	.528	.575
18	.378	.444	.516	.561
19	.369	.433	.503	.549
20	.360	.423	.492	.537
21	.352	.413	.482	.526
22	.344	.404	.472	.515
23	.337	.396	.462	.505

24	.330	.388	.453	.496
25	.323	.381	.445	.487
26	.317	.374	.437	.479
27	.311	.367	.430	.471
28	.306	.361	.423	.463
29	.301	.355	.416	.456
30	.296	.349	.409	.449
35	.275	.325	.381	.418
40	.257	.304	.358	.393
45	.243	.288	.338	.372
50	.231	.273	.322	.354
60	.211	.250	.295	.325
70	.195	.232	.274	.302
80	.183	.217	.256	.283
90	.173	.205	.242	.267
100	.164	.195	.230	.254

SOURCE: Table VII of Fisher and Yates, *Statistical Tables for Biological, Agricultural, and Medical Research*, published by Longman Group, Ltd., London (previously published by Oliver and Boyd, Edinburgh). Used with permission of authors and publishers.

6. CRITICAL VALUES OF *R* IN THE RUNS TEST (P = .05)

N_1 \ N_2	2	3	4	5	6	7	8	9	10	11	12	13	14	15	16	17	18	19	20
4			2																
5		2	2	3															
6		2	3	3	3														
7		2	3	3	4	4													
8	2	2	3	3	4	4	5												
9	2	2	3	4	4	5	5	6											
10	2	3	3	4	5	5	6	6	6										
11	2	3	3	4	5	5	6	6	7	7									
12	2	3	4	4	5	6	6	7	7	8	8								
13	2	3	4	4	5	6	6	7	8	8	9	9							
14	2	3	4	5	5	6	7	7	8	8	9	9	10						
15	2	3	4	5	6	6	7	8	8	9	9	10	10	11					
16	2	3	4	5	6	6	7	8	8	9	10	10	11	11	11				
17	2	3	4	5	6	7	7	8	9	9	10	10	11	11	12	12			
18	2	3	4	5	6	7	8	8	9	10	10	11	11	12	12	13	13		
19	2	3	4	5	6	7	8	8	9	10	10	11	12	12	13	13	14	14	
20	2	3	4	5	6	7	8	9	9	10	11	11	12	12	13	13	14	14	15

For the two-sample runs test any value of *R* which is equal to or less than that shown in the body of the table is significant at the .05 level with direction not predicted, or at the .025 level with direction predicted.

SOURCE: F. S. Swed and C. Eisenhart, "Tables for Testing Randomness of Grouping in a Sequence of Alternatives," *Annals of Mathematical Statistics*, 14: 83–86 (1943). Reprinted with permission.

Glossary

Absolute value. The absolute value of X is the distance of X from 0, or the positive value of X. The absolute value of $(-23) = 23$; the absolute value of $(5) = 5$. Absolute value is indicated as $|-13| = 13$.

Accidental sample. The most convenient or easiest sample to acquire.

Bimodal distribution. Distribution of cases measured on one variable such that two values of the variable occurred with the same frequency in the group and that frequency of occurrence was the largest for the group; that is, two modes, or two most frequently occurring values.

Case. The unit on which variables are measured. Examples: If one measures the number of volumes in branch libraries, the cases would be branch libraries; if one measures salary in a group of librarians, the cases would be librarians.

Coefficient of variation. The ratio of the standard deviation of a group of cases measured on one variable to the mean of the group on that same variable. Useful when comparing two or more groups on the same variable.

Confidence coefficient. The probability level associated with the interval, generally .99 (.95 or .90), which indicates that if we were to repeat our sampling process an infinite number of times, we would expect the population mean to fall within our interval 99 percent of the time, and 99 percent of the time we would expect to get a sample mean within the interval constructed.

Confidence interval. A line segment or interval within which the population parameter and other sample statistics are to be captured.

Confidence limits. The outer values of the confidence interval. If the confidence interval is one concerning sample means, the outer values will be $\overline{X}$ values.

Cross-tabulation table. The simplest form is a 2-by-2 table in which the columns indicate the possible values of one variable and the rows indicate the possible values of a second variable. One group of cases is categorized on both variables and each case is placed into only one of the cells, which is the intersection of one of the rows with one of the columns. The total number of cases for each cell or intersection is recorded and tabulated inside the cell. It represents a bivariate frequency distribution.

Degrees of freedom. For the t test, it is calculated by taking the total sample size and subtracting the number of restrictions placed upon the calculation of the statistic; the number of values that are free to vary in the calculation of the statistic. In the chi-square test it equals the number of rows -1, multiplied by the number of columns -1.

Dependent variable. The variable to be examined and measured.

Descriptive statistics. Statistics used to summarize information about a group of cases measured on one or more variables. No inferences are made from the data. Our purposes are to describe the central tendency and variability of each variable for the total cases and, also, to define any relationships which exist between the variables for our sample.

Estimated standard error. The standard deviation of the sampling distribution of a statistic, which is calculated by using the sample standard deviation rather than the population standard deviation. Because of this estimate of the population standard deviation, the resulting value is an estimated standard error.

Expected frequency. The number of individuals expected in a cell of a cross-tabulation table if the two variables under consideration are totally independent of each other.

Frequency distribution. A table containing two columns: values of the variable and the frequency. Values of the variable are the possible values which the variable takes in our sample. Frequency indicates the number of cases which have the first value of the variable, the number of cases which have the next value of the variable, etc.

Frequency polygon. A graphic illustration of a frequency distribution in two dimensions. The horizontal axis of the graph indicates the values of the variable and the vertical axis indicates the frequency or number of cases in the sample which take on each value. Generally used for data measured at least at the ordinal level.

Histogram. A bar graph used to illustrate a frequency distribution. The horizontal axis indicates the values of the variable and the vertical axis indicates the frequency of occurrence of the values in the sample. The histogram may be used for variables measured at the nominal level.

Hypothesis test. Test of a specific statistical hypothesis concerning population parameters by means of specification of the sampling distribution of the statistic under the statistical hypothesis, selection of a sample size, and selection of a random sample from the population of interest. The statistic of interest is then compared with the sampling distribution of the statistic under the hypothesis to determine its likeliness or unlikeliness and, therefore, the nonrejection or rejection of the statistical hypothesis.

Independent variable. The variable manipulated in order to examine its effect upon the dependent variable. The independent variable generally precedes the dependent variable in time.

Inductive statistics. See inferential statistics.

Inferential statistics. Statistics which employ the logic of induction in order to make generalizations about a population, based upon information gathered from a random sample of cases.

Intercept. In a regression equation involving only one independent variable (X), the intercept equals the value a, where the line crosses the Y axis in the graph.

Interval. A level of measurement in which cases are assigned numbers according to the rules of both nominal and ordinal levels of measurement. In addition, there is a unit of measurement; that is, the distance from 5 to 6 is the same as the distance from 8 to 9 in terms of the amount of the variable which is being measured. An arbitrary 0 point also exists. Example: Individuals may be measured on the variable IQ.

Level of significance. The percentage of area under the sampling distribution which represents what we consider to be the region of rejection, or the values which are unlikely to occur under our null hypothesis. The level of significance is a probability that one will get a value as unlikely or more unlikely than the values in the region of rejection.

Levels of measurement. Traditionally, there are four different levels of measurement: nominal, ordinal, interval, and ratio. The values for a set of cases on one variable correspond with each other according to specific mathematical properties.

Mean. A measure of central tendency which yields one number for one group measured on one variable. The mean indicates the average value of the variable for a group of cases and should be calculated on data measured on at least the interval level. The mean is affected by extreme scores, is associated with the standard deviation, and is calculated by using the formula

$$\overline{X} = \sum_{i=1}^{n} \frac{f_i X_i}{n}$$

Mean absolute deviation. A measure of variability about the mean which indicates the heterogeneity or homogeneity of the group with respect to one variable. Not as useful a measure of variability as the standard deviation since it does not incorporate the same mathematical properties with respect to the normal distribution.

Measurement. A process by which numbers are assigned to represent the different categories that a variable may possess. Example: The variable "sex" may have the assignments male $= 0$, female $= 1$.

Median. A measure of central tendency which is not affected by extreme scores and is the value of the variable associated with the case which has 50 percent of the group above and 50 percent of the group below. The median is calculated on variables that are measured at least at the ordinal level.

Mode. A measure of central tendency which indicates the most frequently occurring value of a variable measured on one group of cases.

n. Used to indicate the number of observations or cases in a sample of data. Example: If we measure 45 librarians on 30 variables of interest to us, the sample size, or n, equals 45.

N. Used to indicate the total number of observations in the population of cases. Example: If one measures a sample of 45 individuals on 30 variables of interest, the sample size n equals 45 but the N size is the total number of librarians we want to generalize about, for instance, the total number in the United States.

Nominal. A level of measurement in which an individual or a case is assigned to a category of a variable. No order is implied by the classification. Examples: Dewey Decimal system, sex, political party affiliation. The number which is assigned to a category of a variable is a "place holder" only, and does not imply that any arithmetical operations can be meaningfully employed. For example, sex: 1 = male, 2 = female. If a person has a 1 on the variable sex, it indicates that the person is a male; it does not indicate order with respect to sex.

Nonprobability sampling. A type of sampling in which every element in the population has an unknown probability of being included in the sample.

Normal distribution. A family of mathematical curves or distributions which are extremely useful in research since so many variables follow a normal distribution in a population. A variable is said to follow a normal distribution if the frequency distribution of that variable closely approximates the shape of a normal distribution. Normal distributions are sometimes referred to as "bell-shaped curves," and their special properties are outlined in chapter 1. If a variable follows a normal distribution or if the sampling distribution of the mean of the variable follows the normal distribution, we can find the percentage of cases that fall between any two values of interest to us by use of appendix table 2.

Null hypothesis. *See* Statistical hypothesis.

Ordinal. A level of measurement in which individuals or cases are assigned to a category of a variable, such that the categories are ordered. Example: An individual may rank-order 10 books according to his preference of the books. In this example the 10 books constitute 10 cases, and the variable is order of preference.

Parameter. A quantitative characteristic of a population, the value of the variable which would be obtained if all members of the population were measured on the variable. It is a fixed number, which is generally unknown.

Pearson product–moment correlation coefficient. A measure of the strength and direction of the linear relationship between two variables in a sample of cases.

Percentage. A proportion multiplied by 100. Example: 77 percent equals 100 × .77 or 77 percent, which indicates 77 percent of the total number of cases or 77 percent of the whole.

Population. The total number of cases of interest, the total number of cases about which a generalization may be made. A population may be countable in number or so large that we might consider it to be infinite and therefore, practically speaking, uncountable.

Population distribution. The frequency distribution of a population of cases on a specific variable of interest. Because we rarely have access to our whole population, the population distribution is generally considered to be a theoretical distribution.

Probability sampling. A type of sample in which every element of the population from which the sample is taken has a known probability of being included in the sample.

Proportion. A fraction, usually expressed as a decimal, in which a small number of cases is compared to the totality of cases. For example: In a library staff of 30 individuals in which 20 are female and 10 are male, the proportion of females is 20 / 30, or .667.

Purposive sample. A nonprobability sample in which the researcher selects cases for analysis based on his or her judgment. Example: including experts in a sample.

Quota. A type of nonprobability sample in which cases are selected until a predetermined number of cases with specific characteristics is included in the sample.

Range. The least useful measure of variability, calculated by subtracting the smallest value of the variable occurring in the sample from the largest value of the variable in the sample.

Ratio. A level of measurement that contains all the properties of the nominal, ordinal, and interval levels of measurement. In addition, there is an absolute 0, which indicates total absence of the property or variable being measured. Examples: number of books in a branch library, salary, etc.

Ratio. The ratio of X to Y is the number of cases with the property X, compared to the number of cases with property Y. Example: If a group contains 10 males and 15 females, the ratio of males to females is 10 to 15, which is expressed as 10 : 15 or 2 : 3.

Regression analysis. A statistical technique that involves one or more independent variables and one dependent variable, in which the form or shape of the relationship among the variables and the relative importance of each X_i in helping to predict Y may be specified.

Rejection region. That region in the sampling distribution of the statistic deemed to consist of unlikely sample statistics under the null hypothesis. Generally, the tail region(s) or sample values which would rarely occur if the statistical hypothesis is true.

Sample. A subset of the total number of cases, a subset of a population of cases.

Sample distribution. The frequency distribution of a sample of cases on a specific variable.

Sample size. The number of cases or observations in the sample, the number of individuals whose variables will be measured.

Sampling distribution. A theoretical distribution of all possible values of a sample statistic which could occur, given that the null hypothesis is true. The statistics are based on independent random samples of a specific n size, all taken from the same population.

Scattergram. A graph which illustrates the "amount" of association between two variables. The horizontal axis represents the values of one variable and the vertical axis represents the values of the second variable. Dots on the graph indicate a pair of values for one case, measured on both variables.

Simple random sampling. A type of probability sample in which every element of the population has an equal chance of being included in the sample.

Slope. In a two-variable regression equation, the value of b equals the slope or steepness of the regression line when graphed.

Standard deviation. A measure of variability for one variable which indicates the homogeneity of the group of cases with respect to the group mean. A large standard deviation indicates that the group of individuals is very different, or heterogeneous, and is widely spread about the group mean. A very small standard deviation indicates that the group is extremely similar or homogeneous, and is closely clustered around the group mean. Most useful when the variable follows a normal distribution.

Standard error. The standard deviation of a sampling distribution of the statistic examined for inferential purposes. Examples: standard error of the mean, standard error of the variance, etc.

Statistic. A quantitative characteristic of a sample. Examples: sample mean, sample r_s, sample standard deviation, sample mode, etc. Statistics vary in value from one sample to another.

Statistical hypothesis. A mathematical statement about the value(s) of a specific population parameter(s) of interest to us. Example: $\mu = 30$; $\rho = .70$; $\sigma = 150$. The statistical hypothesis is assumed to be true until the data indicate otherwise.

Statistical significance. When the sample statistic falls inside the region of rejection in a null hypothesis testing situation, it is statistically significant and the null hypothesis is rejected since the sample statistic seems to fall far away from our hypothesized parameter.

Stratified sampling. A type of probability sampling in which the population is divided into a number of different categories, or strata, and cases are randomly selected (either proportionately or disproportionately) from within the strata. The strata should be categories of a variable which has relevance for the study, in that we would like to include cases from each level of the variable.

Systematic sampling. A type of probability sample in which the first element is selected randomly from the population and elements, or cases, are selected thereafter at a fixed interval, k.

t **test.** A statistical test involving a hypothesis about the population mean of a variable, or a hypothesis about the difference between two population means. The sample size(s) for a *t* test is generally small, and if it is very small ($n =$ less than 30) we must assume that the distribution of the variable in the population is normal.

Unbiased. A statistic is unbiased when the mean of the sampling distribution of the statistic equals the parameter it estimates.

Unimodal distribution. A distribution of cases measured on one variable, such that the group has only one mode or only one value of the variable which occurred most frequently.

Variable. Any item, entity, or phenomenon to be analyzed. A variable is measured for each individual or case according to rules we specify. "Variable" indicates that the values vary from case to case. Examples: sex, IQ, age, number of volumes, rank of preference, etc.

Variance. The standard deviation squared.

Z **score.** A transformation performed on a variable, one case at a time, which results in the cases' having a mean of 0 and a standard deviation of 1. If the variable's distribution is normal, we can use the unit normal distribution (at the end of the text) to find the percentage of cases that fall between any two specified values of our variable.

Answers
to Exercises

Chapter I

1. a. 127 / 903 = 14%
   b. Mean = 154.11; range = 198; s = 65.32

2. a. Mean = 13; median = 14; mode = 14
   b. s = 3.325
   c. $Z = 16 - 13 / 3.325 = .902$
   d.

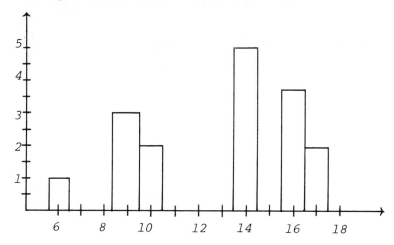

Chapter 3

1. $t = \dfrac{(10{,}750 - 10{,}491)}{\sqrt{\dfrac{150^2}{5} + \dfrac{200^2}{8}}} = \dfrac{259}{\sqrt{9500}} = 2.657$

$df = 5 + 8 - 2 = 11$

Let the significance level equal .05. Critical region: t values smaller than -2.201 or greater than 2.201. Our findings indicate that males and females have different mean salaries.

Chapter 4

1. $r = .56$, a moderate correlation. The following computations emerged in solving this problem.

$$n = 8$$
$$\Sigma Y = 29$$
$$\Sigma Y^2 = 121$$
$$\Sigma X = 620$$
$$\Sigma X^2 = 48{,}600$$
$$\Sigma XY = 2{,}300$$

2. a. $\hat{Y} = -3.776 + .0955X$

   b. $\hat{Y} = -3.776 + .0955(75) = 3.387$

Chapter 5

1. a.

	Full	Associate	Assistant	
User	53 a	33 b	26 c	112
Nonuser	182 d	97 c	71 f	350
	235	130	97	462

CELL

	a	b	c	d	e	f
fe	56.97	31.52	23.52	178.03	98.48	73.48
fo	53	33	26	182	97	71

$\chi^2 = \left(\dfrac{53^2}{56.97} + \dfrac{33^2}{31.52} + \dfrac{26^2}{23.52} + \dfrac{182^2}{178.03} + \dfrac{97^2}{98.48} + \dfrac{71^2}{73.48} \right) - 462$

$\quad = .8 \qquad\qquad df = 2$

Result: there is no relationship between faculty rank and library usage.

b. $C = \sqrt{\dfrac{.8}{.8 + 462}} = .04$

Index

Accidental samples. *See* Non-random samples and sampling

Bias estimator, definition of, 50

C. See Correlation, Pearson's *C*
Central-limit theorem, 48–49
Chebyshev's theorem, 25–26
Chi-square, 78–85
  assumptions for test of interpretation, 78
  expected frequencies, 79–80
Coefficient of variation (V), 19, 21
Confidence interval, 59–60
Confidence limits, 59–60
Contingency coefficient. *See* Correlation, Pearson's *C*
Correlation
  Kendall's *tau*, 91–92
  Pearson's *C*, 86–87
  phi-square (nominal scale), 85–86
  product-moment
    assumptions for independence, 68
    calculation of, 70–71
    significance test for, 71–72

Spearman's r_s (ordinal scale), 89–90
  use with interval scale, 90–91
Critical region, choice of, 46–49, 54–59

Degrees of freedom
  for chi-square, 81
  definition of, 56–57
  difference of means test, 61–63
  t test, when *n* is small, 56–59
Descriptive statistics, 1–27
Difference of means test, 60–64
  assumptions for, 62

Expected frequencies. *See* Chi-square

Frequency, 9
Frequency distribution, 9–11
Frequency polygon, 10–12

Histogram, 12
Hypothesis, null and research, 42
Hypothesis testing, definition of, 42–44

Inductive statistics, definition of, 1, 41
Inferential statistics. *See* Inductive statistics
Interval level of measurement, 4
  tests and measures for, 3

Kendall's *tau*. *See* Correlation, Kendall's *tau*

Levels of measurement, 2–6

Marchant, Maurice, 84, 94
Mean, 13, 15
  difference of means test, 60–64
  hypothesis test for large samples, 53–56
  hypothesis test for small samples, 56–59
Mean deviation, 17
Measures of central tendency. *See* Mean, Median, Mode
Measures of dispersion. *See* Mean deviation, Range, Standard deviation
Median, 14–15
Mode, 15–16

Nominal level of measurement, 2–3
  tests and measures for, 3
Nonparametric tests and measures, 77–78
  nominal scale, 78–87
  ordinal scale, 87–93
Nonprobability samples. *See* Nonrandom samples and sampling
Nonrandom samples and sampling, 37–39
  accidental samples, 38
  purposive samples, 37
  quota samples, 38
Normal distribution, 11–24
  table for, 98
Null hypothesis, definition of, 42–43

Ordinal level of measurement, 3–4
  tests and measures for, 3

Parameter, definition of, 41
Pearson's *C*. *See* Correlation, Pearson's *C*
Pearson's contingency coefficient. *See* Correlation, Pearson's *C*
Pearson's *r* (product-moment correlation), 65–71
  hypothesis test for, 71–72
Percentage, 71
Phi-square. *See* Correlation, phi-square
Population, definition of, 29
Population distribution, definition of, 44
Prediction equations, 73
Probability samples. *See* Random samples and sampling
Product-moment correlation. *See* Correlation, product-moment
Proportion, 6–7
Purposive samples. *See* Nonrandom samples and sampling

Quota samples. *See* Nonrandom samples and sampling

Random numbers, table of, 96
Random samples and sampling, 28–39
  disproportionate stratified, correction for, 36–37
  sample size, 39
  simple random, 30–32
  stratified, 34–36
  systematic, 32–34
Range, 16
Rank-order correlation. *See* Correlation, Kendall's *tau and* Spearman's r_s)
Rate, 8–9
Ratio, 8

Ratio level of measurement, 5–6
  tests and measures for, 3
Regression analysis, 72–75
Runs test, 87–89
  table for, 104

Sample distribution, definition of, 44
Sample size. *See* Random samples
  and sampling
Samples. *See* Random samples and
  sampling
Sampling distribution, definition of,
  44
Scattergram, 66–69
Significance levels. *See* Critical
  region, choice of
Spearman's r_s. *See* Correlation,
  Spearman's r_s
Standard deviation, 17–19, 22–24
  formula for descriptive statistics, 18, 49
  formula for inferential statistics, 50
Standard error
  of the difference between
  means, 61

of the mean, 51
Standard scores. *See* Z scores
Statistical hypothesis, definition of,
  42–43
Stratified sampling. *See* Random
  samples and sampling
Student's *t*. *See* *t* test and distribution
Systematic samples. *See* Random
  samples and sampling

t test and distribution, 56–59
  assumption for single sample
  t test, 57
  See also Difference of means test;
  Mean, hypothesis test for small
  samples

V. *See* Coefficient of variation
Variable, definition of, 1

Wald-Wolfowitz runs test, 87–89

Z scores, 25